HONEY WINES & BEERS

by
CLARA FURNESS

WITH SHORT HISTORICAL NOTES ON
THESE ANCIENT BEVERAGES

Published by
Northern Bee Books
www.northernbeebooks.co.uk

Honey Wines and Beers
ISBN: 978-1-914934-30-8

Photographs by Graham Dyrose, Paul Bond, & Dave Clarke
Cartoons by Alan Ward

This is a facsimile of the original publication of 1988

Published by Northern Bee Books (2022)
Northern Bee Books, Scout Bottom Farm,
Mythomroyd, Hebden Bridge, HX7 5JS (UK)
www.northernbeebooks.co.uk
Tel: +44 (0) 1422 882751

Designed by: Lynnette Busby
www.whatever.graphics

CONTENTS

INTRODUCTION

"The half empty honey pot left out in the rain set man upon the road to alcohol and the illicit still"
"Food in Antiquity" Don and Patricia Brothwell

Most booklets on Mead Making begin with a short survey of the life of the honey bee and the wonders of honey production. This is a booklet for beekeepers, so we can skip that. It is, however, interesting to consider that mead making is probably much older than beekeeping. It must stem from the earliest days of honey hunting. Most of you will be familiar with this reproduction of the rock painting discovered in 1919 in the Arana (spider) cave at Bicorp near Valencia. It is pleasant to think that these stone age ancestors of ours, robbing the nests of wild bees, may by accident have made mankind's first tipple. Unprotected from stings by veil or gloves - and probably by much other protective clothing - this prehistoric hunter would probably make off as quickly as possible with his sweet loot. Some of the precious honey may have dripped from the broken comb on to the rocks below, where it would collect in a hollow. Before the bees had time to reclaim it, perhaps a storm came and diluted it. Wild yeasts, present in all honey (but unable to work because of the high concentration of sugars) would be activated; or those yeasts ever present in the atomsphere would enter the liquid. Nature's alchemy would start. The sugars of the honey would be changed into alcohol. Coming back to collect his esparto grass ladder, our paleolithic bee hunter would quench his thirst at the pool. You can imagine that it would not be long before he was mixing rain and honey on purpose.

Even if our fancies have no foundation in fact, there seems to be no doubt that mead was the first fermented drink enjoyed by man. The honey bee evolved long before Homo Sapiens and there is archaeological evidence of the preparation of honey drinks before the development of wine from grapes.

A nymph of ancient Greek legend named Mellissa is said to have discovered the art of mead making. She it was who fed honey to the god Zeus in the sacred cave of Mount Ida in Crete. It was an ancient belief that those whose lips were touched by a bee in infancy would grow up with the gift of fluent speech or of song. And so it became the custom in many lands to touch a newborn baby's mouth with honey.

The gifts of soothsaying and prophecy were connected with the drinking of nectar as an accompani-

ment to ambrosial food. Nectar was the drink of the gods who were immortal; ambrosia, the food of the gods, comes from two Greek words: a = without and brotos = mortality. In historical times in Greece and Rome, mead was already an ancient drink and that is one of the reasons why it was offered as a libation to the gods.

The mead of Valhalla, the drink of the Norse warrior gods, ran in rivers in that idyllic heaven. The heroes of these legends were huge, heroic and incorruptible. They ate their banquets in the mead hall and drank the foaming liquor from huge mead horns.

Our Roman conquerors grew grapes on our Southern hills, but most of their drinks must have contained honey.

Recent marine archeology shows that wines from Mediterranean countries were imported from earliest times for the use of those rich enough to pay. The native yeoman, however, brewed with native grains and honey.

Honey and mead helped most medicines to go down, their sweet taste ameliorating the bitter herbs which were the cure-alls of those times. It was to

mead they turned when, at the dawn of science, philosophers sought to prolong the span of man's life.

Bacon evolved a "Methusalem Water"; and in 1669 a book was published containing 108 recipes for mead; "The Closet of Sir Kenelme Digby, Knight, Opened". Sir Kenelme was only two when his father, one of the gunpowder plotters, was executed. He became an exile during the Commonwealth and spent a lifetime of colour and adventure all over Europe where he collected his recipes. He had the true character of one whose lips had been anointed with honey. "Digby loved life", says the biographer, "to prolong life, fortify it, clarify it, was his noble pursuit". The diarist Evelyn called him "a teller of strange things", but as Lady Fanshawe said, "He enlarged somewhat more extraordinary stories than might be averred - that was his infirmity". Dr. Harvey, who discovered the circulation of the blood, gave him a recipe for cider. His many recipes for Meathe and Metheglin make diverting reading and I recommend the book. It was reprinted in 1910 and is available from libraries.

The use of beeswax candles on the altars of mediaeval churches led to a tradition of beekeeping amongst certain monastic orders. Beeswax was exacted in large quantity as a tithe by powerful feudal abbots. This kept the practice of apiculture active. As a sideline, the practice of mead making became the expert occupation of many cellarers serving religious communities. Most of the famous liqueurs of France were first produced by priestly orders and still bear their names. The recipes are jealously guarded secrets. There must be many mead recipes in the archives of monastic houses which rose from the ruins of the dissolution. So far I have failed to find any. Scottish drambuie, traditionally based on heather honey, is made from a secret formula.

Mead has always been looked upon as a love potion. It was poured as a libation at the altar of Aphrodite, the goddess of love. It was carried in wine skins to Bacchus or Dionysus, the god of wine. In all Northern lands, where the honey bee is native, mead was the traditional drink at weddings. There are stories of vast quantities being drunk at royal nuptials. In those far off days when people were usually hungry, you can imagine the sudden surge of energy and well-being occasioned by eating honey and drinking honey wines. No wonder they were looked upon as the food and drink of romance and were allowed for a full month to the betrothed couple. In this way they hoped for the early conception of the much desired child.

All sorts of customs surrounded the honeymoon. Honey was smeared on the forehead and lips of the bridal couple to sweeten the words they would say to each other. Mothers offered mead to their new sons in law. Fathers decorated the beehives and brides decked them with ribbons and flowers. The origin of the superstition that a bride should be carried over the threshold of her new home stems from the custom of daubing step and lintel with honey to ensure a sweet marriage free from strife.

Mead makes a good wine for a modern wedding and a sparkling mead can replace the champagne very well. Mead and honey wines come in a wide variety and there is one suitable for every event which calls for celebration.

The beekeeper - meadmaker who keeps up a regular supply of these delightful drinks has no need to reserve them for special occasions. He can take a glass with every meal, making his humble daily fare into a banquet. He can regale his visitors with honeyed wine, and invigorate the bee talk. The meetings of his beekeepers' association can become minor celebrations. He can whet his whistle, warm his nose, calm his nerves, dampen his smoker, toast the bees, drown his stings, placate his spouse and always truly say: "a little of what you fancy does you good".

Mead settles your tummy, helps you to sleep and is a fine tonic for those poor folk who suffer from constipation. And even if you are up against the odd rechabite, you can truly say "It's purely medicinal you know".

TYPES OF MEAD

"Variety's the very spice of life,
That gives it all its flavour" - W Cowper 1731-1800

Mead, says the concise Oxford dictionary, is "alcoholic liquor of fermented honey and water". The word has common derivatives in old English, Dutch, German, Sanskrit, Greek and Latin and seems to have roots in the word used both for honey and the drinks made from it. Sir Kenelme Digby uses the names Meath, Meathe, Metheglin, Hydromel equally for drinks made with water, honey and a large variety of spices and herbs. Some of the herbs are still in use in our kitchens, but many are wild or cultivated plants long ago discarded as culinary aids. He gives no recipes for what we would consider a true mead - a wine or beer made only from honey, water and yeast and depending for flavour only on the honey. In those days there was little distinction between the kitchen and the apothecary's "elaboratory" and even queens and great ladies and gentlemen turned their hands to making medicinal decoctions along with their daily fare. Most of the old recipes for meath have some medicinal intent. The Celtic word Metheglin derives from meddyg = healing and llyn = liquor. It obviously has the same etymology as medicine.

Mead disappeared from common use in the centuries which saw the rapid expansion of population and growth of towns. The well-to-do favoured the cheap wine imported first from France and later from Germany, Italy Spain and North African States. The drink of the working man was ale or beer.

Only the country housewife who still tended her skeps kept alive the old skill of mead making. She probably looked upon her drinks as medicinal and it is certain that they were drunk almost as soon as they were made, with little regard for clarity or maturity. Only the aristocracy and senior clergy were honoured with clear, mature "old" mead or ale. The common lay brother and the peasantry drank their still fermenting meathe or barley broth. In this way they received the health giving yeasts and unfermented honey sugars to counteract the skin eruptions and scurvy scourges of their age. They also precluded the development of diseases in their meads and honey ales.

Let us remember that it is only just over a hundred years since Pasteur disclosed the secrets of fermentation. Before that, spoilage of wine and beer during manufacture was common. It was thought wise to drink the young brews before they fell prey to all the diseases which their makers feared.. Spoiled wines or beers and the weaker drinks made from the spent grains or second washings of the honeycombs were the rations of the commonalty or poorer folk. These were sweetened with honey and flavoured with herbs to make them palatable.

The highly successful temperance movement, part of the puritanical development following the Commonwealth, was strengthened by the unhappy effects of cheap gin in the late 18th and early 19th centuries. It spread throughout the land and was especially successful in the nonconformist industrial North. Its commendable preaching of temperance changed into an implacable movement for total abstinence, and health giving home ale and mead brewing was lost in its wake. It was even for a time in this century illegal to brew at home. Happily for us, these prohibitions are no more. We can make as much home brew as we wish, but only for our own consumption. It is still illegal to sell home made beers, wines and meads or even to offer them for money raising events such as raffles.

In this modern, sophisticated age, even the least of us eat and drink like the aristocracy of old. Diseases of wine and ale are fully understood and guarded against, and we ourselves have a much increased expectancy of age. We take advantage of scientific progress, clinical cleanliness and experience of drinking good, imported wines. We can consequently produce for ourselves a mature and sophisticated drink.

Mead is not wine. Wine is, strictly, prepared only from the juice of grapes, but this word is used by all who enjoy mead and, for convenience, I too shall use it. There has been a phenomenal growth in home winemaking in the last quarter of a century. The reasons for this are many, but foreign travel, improved standards of living, more leisure and a blossoming of the philosophy of self sufficiency have contributed. I also think that a commendable temperance has taken the place of the uncompromising teetotalism of our grandparents' lifetime. Home made mead is not to provide a sordid intoxication – it is to be a gracious accompaniment to a home cooked meal and a shared conviviality of our beekeeping hospitality.

The spread of home winemaking shops has provided an abundance of materials, equipment and simple scientific instruments for the more knowledgeable participant. Many amateur winemakers stem from families who never gave up the age old practice of brewing. I am one of them. My mother made her own herbal medicines and brewed her own earthy beers and wines as naturally as she kneaded her own bread, baked cakes and scones and prepared jams and pickles in season. But those simple brews would be scorned by the new winemaking fraternity.

When we produce our blackberry and apple wines, our elderflower champagnes and our dandelion beers we are making old country brews, but with modern techniques. In the early days of winemaking from grape concentrates we bought cans labelled "Chablis type", "Bordeaux type" etc. This is now illegal and the concentrate is described more simply as "red dry", "sweet white" or just indicates the country of origin. We are still, however, making an imitation of the true wine produced from fresh grape pressings. Not so with mead. When we prepare our honey musts we are imitating nothing. We are making real mead, the good tipple of our native land.

Thomas Hardy in one of his Wessex Tales, "The Three Strangers", has his second stranger, drinking from a huge vessel of brown ware inscribed with the legend "Ther is no fun until I cum" remarks, "I knew it. When I walked up your garden before coming in and saw the hives all of a row, I said to myself, "Where there's bees, there's honey, and where there's honey, there's mead". I am assuming that many of the

readers of this booklet will be beekeepers. How many of these can always, like Shepherd Fennel of "The Three Strangers" be sure of having a glass of mead to offer to every visitor who sees his or her hives? And where is the beekeeper who sets a glass of mead by his plate each time he dines? There are some and I am one of them. Even if you have to buy your honey, your mead will not cost more than a grape concentrate wine. And I guarantee you'll like it just as much or even better. Many country winemakers and mead makers tend to stick faithfully to old customs, and instructions appear in journals which disregard recent developments in home winemaking. I hope the advice in this book will combine much that is new, convenient and simple with that which is traditional. Together they can produce a drink to grace our tables, delight our friends and maybe raise a trophy or two to honour both the maker and his or her local beekeepers' association.

A LIST FOR REFERENCE

"He sent hir piment, meeth and spyced ale,
And wafres pyping hot out of the glade,
And for she was of toune, he profered mede"- Chaucer
The Miller's Tale

This is a list which is not necessarily based on traditional terms, but which has recently been accepted by beekeepers and winemakers. In the old herbals and cookery books all honey based wines were referred to as mead, mede, meeth, meathe, metheglin. Most of them included spices and herbs of a bewildering variety. None was limited to honey + water + yeast, the mead of the modern purist.

HYDROMEL

Strictly water + honey.

The name is given in France to a wine made of grape juice enriched with honey. In fact French writers rejoice in the fact that the grape and honey harvests coincide and so facilitate the making of hydromel. You should buy a bottle when you visit French apiaries. Unfortunately there is little future for hydromel except as a tourist attraction, as it costs more to produce than does a respectable table wine. Sir Kenelme Digby used the term when he made hydromel for the Queen Mother, Henrietta Maria, exiled in France after

the execution of her husband King Charles I. I use the word for unfermented honey drinks which I include. Children may enjoy a glass at table with their mead drinking elders. Hydromels may be offered to those of our guests who prefer or are obliged to avoid alcohol.

MEAD

The currently accepted definition for a drink made from honey, water and yeast with no additives than those necessary to avoid spoilage or assist fermentation. These include acids, nutrients and tannin. It certainly refers to the drink you can enter in the mead class at a honey show. Mead may be of every grade of sweetness from completely dry to very sweet.

Dry mead is a light table wine, a suitable accompaniment to meals. It should be delicate and may be slightly effervescent. In this case it is a delightful aperitif. It should have no residual sugar. Its alcohol content should be low: 9% to 11%.

Sweet mead, often called sack mead, because in tudor times it was compared with the Spanish sack wines, sherries and Madeiras, is a dessert or social wine. Some of the honey sugars remain unfermented. It has more body and can be expected to carry a higher proportion of alcohol: 12% to 15%. It is drunk as an appetiser by those who would choose a sweet sherry. It accompanies the sweet course of a meal or makes a good party drink. It may be used in a long drink by mixing it with soda or tonic or bitter lemon. Add a slice of fresh lemon or cucumber or a sprig of borage to be adventurous. Sweet mead must not be so sweet that the vinosity is lost.

Medium mead, somewhere in between, can be of great variety, ranging from medium dry to medium sweet. This is probably the best of meads, sufficiently dry for your palate, but giving a little sweetness to show its origin in honey. This wine is best used as Germans use the lovely Hocks, Mosels and Nahe wines – as an

accompaniment to an evening's conversation or cards. Don't let the alcohol exceed 12%.

MELOMEL

This is a modern word and denotes a wine made with fruit juices and honey. The finished drink contains both flavours — that of the fruit and that of the honey. I look upon them as the most important products of the beekeeper – winemaker. Many melomels have special names, some with a long history, some traditional in certain localities, some recent inventions.

Apples & honey	cyser
Grapes & honey	pyment or clarre
Red currants & honey	red mead
Black currants & honey	black mead
Mulberries & honey	morat or alicante wine
Bilberries & honey	myritis
Rosepetals & honey	rhodomel
Hops & honey	miodomel

METHEGLIN

In recent times this word is used to denote a mead or melomel flavoured with herbs and/or spices. These make wonderful aperitifs. In Winter, try serving as hot toddies to welcome guests at a party.

HIPPOCRAS

Any mead or melomel used as a vehicle for medicinal herbs and spices. It is possibly more highly flavoured than metheglin. A bottle of hippocras with fresh sprigs of herbs standing in the crystal clear mead is a perfect gift for the recipient who has everything!

MULSUM

This was drunk by the Roman occupation troops. It was imported wine made into a long, sweet drink with honey and water. Its name led to "mulled wine", a feature of Christmas parties, hot spiced wine sweetened with honey.

BRAGGOT or BRACKET

A mixture of ale or beer and honey and often of spices. The fourth Sunday in Lent, commonly called "Mothering Sunday" used to have the name of "Bracket Sunday" in Lancashire. This enriched beer was drunk when serving men and maids visited their own homes to take posies of wild flowers to their mothers.

INSTANT MEAD

"Honey gives life to Wine when it is flat" - Purchas

An amateur winemaker must be patient. It takes at least six months to have anything drinkable and a year is better. The old mead makers matured their brew in oak casks for ten years, and this is the advice given by Brother Adam. With modern aids to fermentation, using glass fermenting jars, our mead is probably at its best in its third year. You will need something good to be drinking while you are waiting. Otherwise you may be like a few of my wine-making friends who never actually have anything to drink. It all disappears in "Tasters" long before it is ready.

You will require:
1 gallon and 1 half gallon container
Two fermentation locks to fit these vessels
sterilised wine bottles and corks
Honey is warmed by standing the jar in warm water. It is warmed so that it mixes more easily. If you are in one of the cider producing counties, buy a gallon of rough cider, otherwise settle for a gallon of commercial stuff from your off licence. Buy also a large bottle of white wine at the bargain counter. Warm ½ - 1kg (1-2lbs) honey (depending on whether your taste runs to dry or sweet wines) and mix this with some of the cider. Stir thoroughly. Crush a campden tablet and dissolve it in about a dessertspoonful of warm water. Stir it into the honey/cider mixture. Remove sufficient cider to make room to add the mixture to your cider and fill to within 3cm (1 inch) of the top. Rotate the jar to ensure thorough distribution. Fit the fermentation lock and keep it at room temperature for at least three weeks, when it should be clear. Siphon into sterilised bottles. This is your INSTANT CYSER

Follow this up with INSTANT PYMENT

Mix the spare cider with your wine and stir in 250 -450 g (½ - 1 lb) warmed honey. Put the mixture into your half gallon jar. Fill up with wine (home made or commercial) cider or fruit juice. Fit a fermentation lock and keep for three or four weeks.

If these wines do not clear, remove to a very cold place for a few days. Sipohn your instant pyment into sterile bottles. Honey contains minute particles of wax and pollens and often forms a very light deposit, so siphon with care. You can add bentonite which will help to firm the lees.

WE WANT TO DRINK IT OURSELVES, NOT BEQUEATH IT TO OUR GRANDCHILDREN.

DO NOT OPEN UNTIL 2000 A.D.

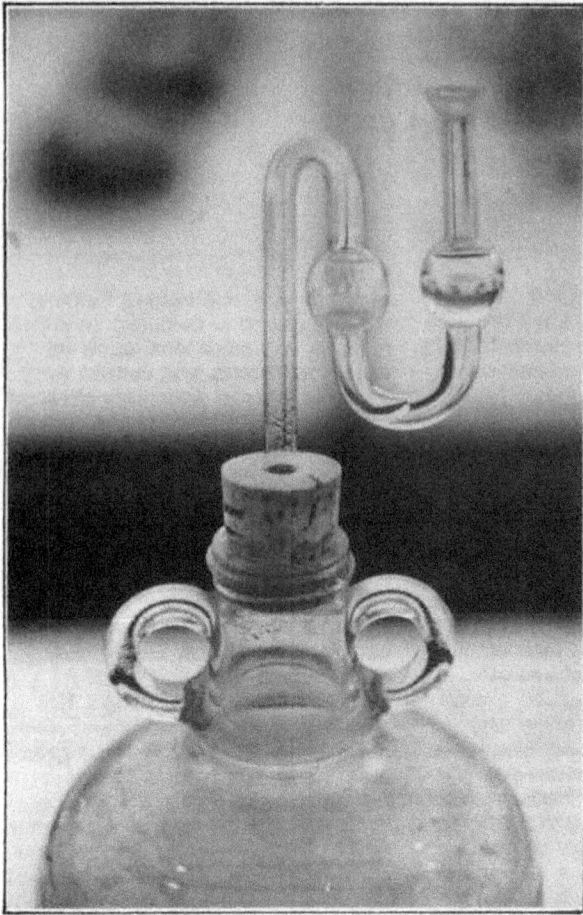

Glass fermentation lock

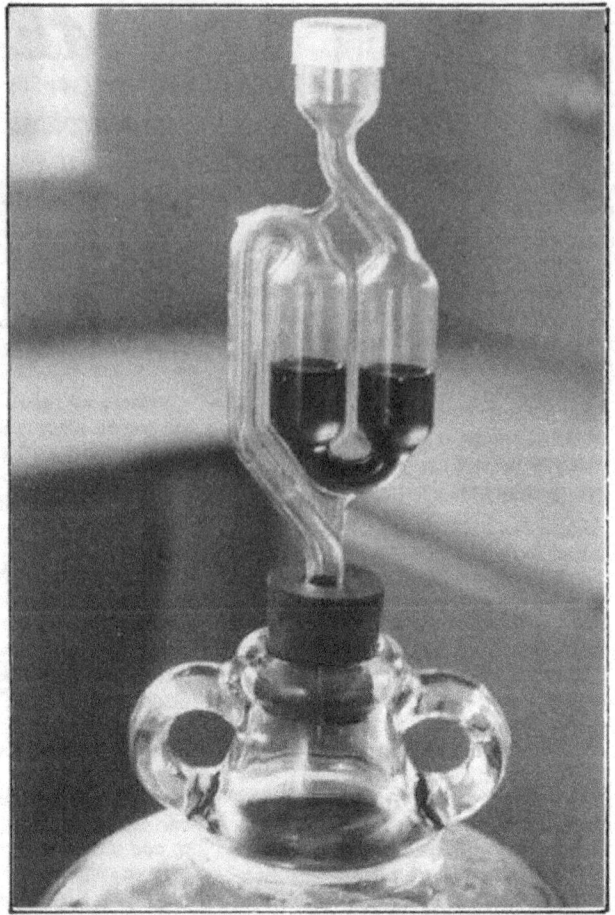

Plastic fermentation lock

Essential tools

Demi-john with cork and simple fermentation lock

MEAD

"Mead ten or twelve years old is a most sovereign and a pleasant remedy for many diseases"
Galen (quoted by Purchas)

There is no legal definition for mead or, indeed for any wine drunk in the U.K. I think that, since we enter mead for competition in honey shows, it would be in the interest of beekeepers to establish a code of practice. Until we do, it is useful to agree on the terms we use.

DRY or SWEET?

Most honey shows have these two classes in their schedules, but it is difficult to define the classes accurately. We have no yardstick comparable to honey grading glasses which can help the competitor to decide where to place his mead. I think it would be helpful if we had a third class for those delightful meads which are neither dry nor sweet and could be classed as medium. Honey is so obviously sweet that mead really ought to preserve some of that sweetness in its final flavour. Yet if the schedule says dry, surely it ought to mean what it says. In a dry mead, ALL the sugars of the honey should have been converted into alcohol, and the flavour should come only from those trace substances which give that honey its individuality. Dry mead is an acquired taste. Few people like it at first, but its devotees consider it a wonderful wine, especially to accompany fish, shellfish and white meats.

As it is, completely dry and medium-dry meads are acceptable in a dry class. Very sweet and medium-sweet meads are together in the sweet class.

In the past mead has usually been too sweet. Many people will not even try mead because they suspect it will be far too sweet and cloying. How many of us have offered mead to a visitor who has refused it because he "doesn't like sweet wines", but who accepts a sherry which is actually sweeter? Sweet mead should not be so heavy with unfermented honey that its vinous quality is lost. The balance, roundness, astringency should be discernible. Judges are usually tolerant of meads on the fringes of the range between dry and sweet. My experience, however, leads me to believe that meads I would consider medium-sweet can win a dry class. I would dearly like to see a class in the major shows for truly dry mead.

You have only to read through the journals to see how often contributors enquire about standards for mead making and judging. In recent years wine judges have been employed at most shows and standards are high.

The London National Honey Show schedule states: "No alcohol may be added to wine, metheglin or melomel, nor may alcohol or flavouring be added to mead, **but additions such as acids, nutrients and tannin may be used.** Brother Adam (whose reprint from "Bee World" 1953 pp 149-156, available from I.B.R.A. is probably the most popular leaflet on mead) insists "no chemicals should be added to the liquor... and it must be stored in wood a full seven years before it is bottled ... (then) it will surpass the finest wines produced from the juice of the grape."

Well ... sherry and brandy casks are well nigh unobtainable. Even whisky distillers who mature their liquor in these casks cannot get enough. This booklet is for the winemaker who plans to make mead by the 5 litre or 1 gallon, using the facilities of an ordinary kitchen. We also want to drink it ourselves and not bequeath it, untasted, to our grandchildren. So we must have additives.

The basic constituents of mead are:
1. Honey
2. Water
3. Yeast and yeast nutrients
4. Acid
5. Tannin

Let us consider each item in turn.

---

1. HONEY FOR MEAD

"And who knows honey sweet, who tastes it not?
Nature demands; Deny? It is a crime" - J B Haurean Mss in Vatican
trans. by Helen Waddell

Most publications about wine making stress that only first class ingredients produce good results. If we apply this argument to honey-for-mead we must first ask, what do you mean by good honey? In my opinion any honey can be used. Even that perfectionist Bro. Adam says, "It is one way of using the honey which would otherwise be thrown to waste".

Honey good on bread and butter is not necessarily a good mead honey. Some basic tastes, masked by the sugars of honey combine to a strange piquancy. When all the sugar is fermented, what remains may not be a good flavour for wine. I would recommend you to use any honey you can lay your hands on.

Some people wash their cappings during extraction of wax for the show bench. These washings are the traditional source of the liquor for mead. Heather press washings are a most enviable source. Mrs. Fennell in Hardy's "The Three Strangers" made mead of two kinds, one a small mead from comb washings and the other a strong mead from honey itself. Ale yeast would be used for both. "It is trouble enough to make - and really I hardly think we shall make any more. For honey sells well and we ourselves can make shift with a drop o' small mead and metheglin for common use from the comb washings."

Many beekeepers have a few pounds of honey which have neither the appearance nor the consistency they like. These could be earmarked for mead. The very dark honeydew honeys which some of us love, but which may not be popular with some of our customers, often make excellent mead. The dark particles in suspension fall out and join the lees in the bottom of the fermentation vessel, leaving a golden liquid which bears no relation to the dark opaque honey.

If you are making a dry mead, you will be fermenting out all the sugars. The residual taste will be only of those trace substances which give the honey its unique flavour. It is not easy to decide what those obscure flavourings are when you taste the original honey. The strong, sweet taste of the sugars masks the underlying savours. Still a honey you like will probably produce a mead you like.

In the case of sweet mead, you are left with quite a taste of the sugars in the honey as well as of its unique aromatics. You can choose a stronger tasting honey here quite safely.

When making melomels, it is good to experiment with different honeys. The double bouquet and savour of the opposing tastes of the chosen fruit and honey, are something to test out and balance.

Nowadays, as in Hardy's time, honey fetches a good price, so it is unrealistic to expect beekeepers to select their "best" honey for mead making, except, perhaps, for those brews aimed at the show bench. I am willing to make mead out of any "unsaleable" honey. The last few pounds from the storage tank, the uncapped honey drawn off before the real extraction, any residue from last season and even honey from the old brood combs all go into my honey wines. Even honey which has begun to ferment is good for mead. It may not produce the very best of brew, but do not despair. Later on I hope to suggest ways of using our less attractive batches of mead.

2. WATER FOR MEAD

"And Noah he often said to his wife
As they sat down to dine,
'I don't care where the water goes,
If it doesn't get into the wine' ". - G K Chesterton
The Ballad of Wine and Water

In 1639 John Taylor, known as the water poet, published a book called "Drinks and Welcome" which included "a description of all sorts of waters from the ocean sea to the teares of a woman". We won't go so far. We'll just try to find out the best water to put with our honey to make mead. Though we may like to read that even John Taylor used the stuff for metheglin.

'Thus water boyles, parboyles and mundifies,
Cleares, cleanses, clarifies and purifies.
But as it purges us from filth and stincke
We must remember that it makes us drinke,
Metheglin, Bragget, Beere and headstrong Ale,
(That can put colour in a visage pale)
By which means many brewers are grown rich,
And in estates may soare a lofty pitch."

Shortly after Taylor published his book the brewers of London were forbidden to use conduit water. This was water piped into London from Hertfordshire because the normal supply, Thames water, was so polluted. If brewers helped themselves, it prevented ordinary folk from getting their share. Consequently mead and ale makers were advised to use clean springs and rain butts. Good water was a luxury then.

Mead recipes have come down from those times. I have seen them slavishly copied in quite recent articles; and there are still people advocating any water except that obtained from the mains supply. Many even accuse tap water of containing deleterious substances such as rust, chlorine and soil. The mediaeval prohibition has been turned topsy turvy.

In fact much of the literature on mead making followed the example of that on beekeeping and repeated the wisdom of the ancients. Galen recommended fountain water, but with much boiling of the must. Pliny's choice was rain water kept for five years, or rain water just after it had fallen, boiled down to one third.

What, I ask, is the advantage of progress? Our water authorities engage experts to supervise the quality of our tap water. This precious commodity, over used and under valued, has been most thoroughly tested and sterilised before being put into supply. Lead pipes are no more and all we have to do is turn the tap and drink it. It comes out clean, sterile and oxygenated. It also contains small amounts of sodium, calcium and magnesium salts, e.g. sulphates, chlorides, phosphates and bicarbonates valuable in the formation of desirable products of fermentation. Whilst mains water may, and probably does contain some of these valuable trace compounds, the water works chemist ensures that these (and any other substances) are not present to the extent where they might be harmful. If you choose to go to your rain butt, the water may have run over lead guttering or been in contact with galvanised pipes or tank, and consequently could contain excessive amounts of lead or zinc. It will certainly have washed the soiling from your roofs.

Have faith in your mains supply - and there is no need to boil it.

THE MUST

"For drink the grape She crushes, unoffensive must and meaths
From many a berry and from sweet kernels press'd
She tempers dulcet creams" - John Milton

When the water and honey are mixed together the resultant solution is called the must. Neither water nor honey will ferment on its own, but when mixed the dilute sugars are attractive to yeasts and an undesirable ferment will start if you do not quickly institute your intentional ferment.

There are several wild yeasts present in honey, and the surrounding air contains wild yeats cells as well as other organisms which will spoil any wine-must left exposed to them. So do not mix honey and water in advance of your mead making and keep it closely covered at all times. Every amateur winemaking book gives full advice on the sterilisation of musts. There are two

methods - boiling and treatment with Campden tablets - sodium metabisulphite. All the old recipes advocate lengthy boiling, often for half an hour, with vigorous removal of scum. Since many of the flavours for which mead is renowned are volatile, lengthy boiling will adversely affect the bouquet and taste. If you must boil, bring the liquid rapidly to boiling point and cool it quickly. Cover at all times, except when just on the boil, when it foams and would boil over. If using sulphite two Campden tablets crushed and mixed with a little warm water should be added to each 5 litres (1 gallon).

Prof. R. Morse in an unpublished thesis from Cornell University (1953) "The Fermentation of Diluted Honey" described experiments in the sterilisation of honey musts. He found that the natural yeasts present in honey quickly succumbed to the wine yeast culture used so there was no need to boil or treat with sulphite. Since all my winemaking involved one of these methods of sterilisation, I tried fermenting untreated musts with some trepidation, but I experienced no failure. I now do this regularly, except in the case of honey already fermenting.

Of course sterilisation of all equipment used and the exclusion of air was observed at all times.

YEAST FOR MEAD

Fairy to Puck "...Are you not he
That fright the maidens of the villagery;
Skim milk and sometimes labour in the quern,
And bootless make the breathless housewife churn;
And sometime make the drink to bear no barm?" - Shakespeare

To most people over middle age the word yeast conjures up the moist, crumbly mass used on baking days when they were children. In my Yorkshire childhood, the word was pronounced 'yest' and this is how it was spelled in many ancient cook books. To younger folk it will probably recall school lessons in biology or botany and some ideas about beer and wine fermentation and growth without oxygen. To amateur winemakers it is a subject of bewildering complexity.

To repeat: There are wild yeasts present in honey, but unable to develop because of the high sugar concentration. There are also yeasts present in the air all around us. As soon as a weak solution of honey and water is exposed, these yeasts enter the must and begin to multiply. Our wild yeasts tend to have a very low alcohol tolerance. They die off as soon as 4% to 5% alcohol has been produced. This would result in a mead of beer strength. That is why the mead of our Northern ancestors was a long drink which had to be consumed quickly before it deteriorated. They had no knowledge of yeasts and had to take what came from their native air.

The biggest difficulty in the past was for beekeepers to find a good yeast which would ferment out well, give

about 11% alcohol and a firm deposit or lees. The best technique was to crush about a pound of fresh grapes and mix these into the honey must. If you were lucky, the natural yeast on the grape skins would take over and give a reliable wine-style fermentation. We have no need of these chancy methods. The home winemaking movement has solved all our problems. You can buy wine yeasts in dried form in packets for one small batch or boxes for bulk mead making. You can buy it in liquid form in small hermetically sealed plastic bottles. It comes by the name of all the well-known (grape) wine types.

This is where knowledge of commercial wine gives you confidence. You select the wine type which your idealised mead is going to emulate and buy the yeast cultured from that grape. This is obvious advice for the home winemaker who may wish to emulate a Sauternes with his apple wine or a Burgundy with his elderberries and blackberries.

But the mead maker is imitating nothing - he is making a sophisticated version of a true native wine. It is therefore best to consider the purpose of the mead we are making. A light, dry mead will need a yeast cultured from a light, dry wine such or Hock or Chablis. A medium sweet mead will call for a

Sauternes yeast or one of the Moselle types, Piesporter or Bernkasteler. Sweet mead is better made with a yeast with a high alcohol tolerance - sherry, port, Madeira. I like Tokay yeast, though this is a wine fortified not with brandy but with extra sweet grape juice.

Most manufacturers of home winemaking materials sell a general purpose yeast. There has been much competition to produce a reliable, quick acting, trouble free preparation. It can be added directly to the must. Beginners in mead making should start with one of these yeasts. Take advice from the proprietor of your home wine shop.

A mead yeast was produced in the 1970's and can still be bought. I have not had great success with this yeast which gives a flocculent deposit.

As you advance with your mead making, try a liquid wine culture. They are more expensive and more complicated in use but I think you will appreciate the finer qualities of your finished mead. The yeast is allowed to develop slowly in a small amount of must for a few days. Only when this culture is well established, should you add it to the bulk of the must. Full instructions are included in every yeast you buy.

YEAST GROWTH

"I have discovered an infusorial animalcule which is one of the principal agents of putrifaction and which possess the singular faculty, previously unknown to the natural sciences, of living without air" - Louis Pasteur
Letter to the Minister of Education 1862

The yeast you buy is in a state of rest. When yeast cells enter a nourishing liquid like a honey must, they begin to grow. The cells bud and separate, making new cells. For this growth they need air, so this first part of the process is called the aerobic fermentation.

When the yeast has grown sufficiently, air should be excluded. This may occur naturally by a cushion of carbon dioxide gas forming over the must. We help to exclude air by means of a fermentation lock. The yeast cells continue to multiply until they have used up all the air present in the must. After that, they act on the various sugars present and convert these into alcohol and carbon dioxide gas. The alcohol forms your wine and the gas escapes in the bubbles

This is called the anaerobic fermentation. Be careful that this vital part of your mead making remains anaerobic by checking the efficiency of your fermentation lock.

YEAST NUTRIENTS

In order to do its work yeast, like every other living organism needs food. It depends on sugar for its energy source, but also needs other nutrients to assist metabolism and healthy growth. This may be compared to the way plants need fertiliser to grow to their full potential. Yeast nutrient salts and plant fertiliser contain the same "golden tripod" of elements: nitrogen, phosphorus and potassium.

Grape juice is the perfect wine must, containing in itself all the constituents required for fermentation. Even its own yeast occurs on its own skins. Other fruit juices contain certain nutrients, but not enough to support complete fermentation. Nutrient salts are added to these musts. Honey is deficient in most of these growth factors and so mead needs a much bigger helping hand.

5 litres (1 gallon) of mead will require 8g (¼oz) of nutrient salts.

½ teaspoon (1 level teaspoon) of nutrient salts weighs approx. 5g.

Yeast nutrient is a mixture of mineral salts prepared in the proper balance for the home winemaking trade. We do not need to trouble ourselves with details of their composition.

Vitamin B also helps to promote a healthy ferment. Use one 3mg vitamin B tablet to each gallon of mead.

Marmite is a concentrated extract of dead yeast, rich in vitamin B; about a quarter of a teaspoonful of this may be used instead.

Mead makers in soft water districts could also add a pinch of Epsom salts (magnesium sulphate).

If you wish to prolong a fermentation and produce a mead very rich in alcohol, you can add honey in small quantities (say 100g (3-4 oz) at a time and feed with tiny additions of vitamin B. These meads make good punches. They also pull a good punch and should be drunk with caution.

ACIDS FOR MEAD

*"For he on honeydew hath fed
And drunk the milk of paradise"* - S T Coleridge
Khubla Khan

Honey diluted with water will ferment with or without the addition of yeast, acid or anything else. But what you get will be barely drinkable. Many types of honey are slightly acid in character, but not sufficiently to support a good fermentation. It is vital to the final flavour of your mead to begin with a good balance of acid in your must. As salt brings out the flavour of food, so the acid added to mead accents the subtle flavour of the honey in the final wine. It also contributes to the work of the yeast, producing by-products of fermentation with acceptable qualities.

If the yeast works with too little acid present, by-products with a mousey or fusty smell are produced. The taste also

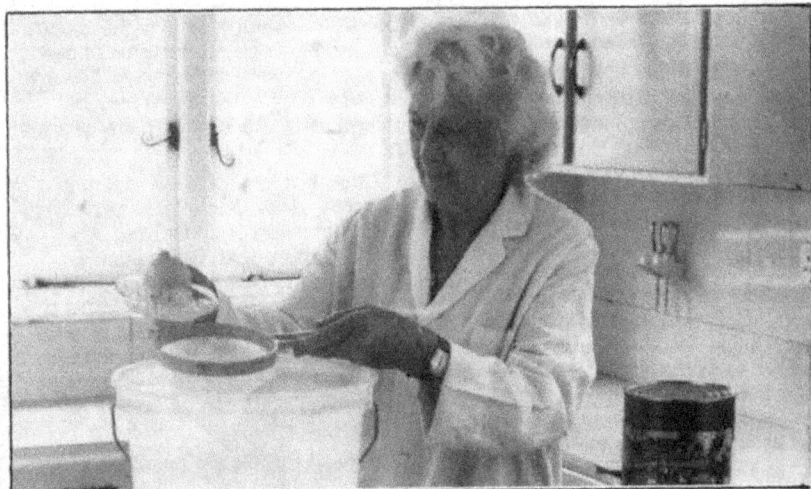

Natural additives

Natural additives. Juicing apples by electricity

is affected and there is a suspicion of a musty or medicinal flavour. Do not use an excess of acid with this in mind, thinking to rectify it later. Mead is a subtle, living thing. Good balance pays.

Now, if we are making mead for our own table, it does not matter in what form we add our acid, so long as we have sufficient to provide desirable products of fermentation. If we are making it for the show bench, we must conform to the schedule. That of the London National Honey Show states: "additions such as acids...may be used." Unfortunately it does not say which acids. Many fruit juices are acid and will raise the acidity of a honey-must if added to it. I imagine that almost every mead which wins a prize has its acid content in the form of lemon juice. I have asked many

prizewinners what acid they added and most have replied, "lemon juice". I believe beekeepers all over the country add lemon juice and a little bit of peel or zest to most of their meads. The acid in lemons is citric acid and may be bought in crystal form from any chemist or home wine store.

Other fruit may be used to add acid to honey wines. Apple juice, especially that from sharp or underripe fruit, contains malic acid. Grapes contain (according to type and maturity) varying amounts of tartaric acid. This forms a crystalline crust on the insides of wine casks and is cracked off after the cask has been emptied of the finished wine. Both these acids can be bought as crystals.

If mead makers are permitted to add

citric acid as fresh lemon juice, would it not also be permissible to add malic acid as apple juice and tartaric acid as grape juice?

It is now widely accepted that if honey is mixed with fruit juices, the resultant wine must be classed as a melomel. If, however, only sufficient juices are added to provide adequate acid fermentation, can we not class the resultant drink as mead? I like to think so. I still, however, label my meads as having "competition additives" or "natural additives".

To add acids in crystal form, buy citric, malic and tartaric acid. Mix them together and use about 15g (about ½ oz) (One rounded teaspoon holds about 5g of mikxed acids.) You can experiment with different proportions of the three acids.

MELOMEL

"Yet oft have we feasted in days of old
And hazel mead drank from cups of gold" - P J Joyce Old Celtic Romance
Quoted by H Ransome in The Sacred Bee

I am including the making of fruit-honey wines or melomels along with the discussion of acids, as natural acids are added this way. The recommended mixed acids must be reduced or omitted altogether in proportion to the acidity of the fruit employed. Readers should remember here, as elsewhere, that recipes are not precise or sacrosanct instructions to be slavishly followed. They are, rather, guidelines for you to use whilst inexperienced, but to develop and adapt to other ingredients.

Melomels may be looked upon as mead enhanced or flavoured with fruit or as fruit wines with the added aroma and savour of honey. It does not matter at all what quantity of fruit juices are employed and many mixtures are attractive.

Pure grape juice with a little honey gives a unique wine with a lingering reminder of nectar. It rejoices in the name of pyment. The word derives from pigment, since the pale wine was originally coloured or pigmented by the honey. Apple juice with honey gives a sharp, truly English wine named cyser. Citrus fruits make a quick maturing wine and are a good choice for the beginner to try. Tinned fruit and fruit juices as well as fresh fruits in season, wild and cultivated can go into honey wines.

The fruit used for melomels may contribute nutrients and tannin, but high tannin fruits such as elderberries make such distinctive wines it is a shame to combine them with honey.

Grape juice remains the perfect wine base. It is available in partially dehydrated form in an enormous variety. Since it is concentrated by freezing and not by application of heat, it is just about identical with fresh pressings once you have reconstituted it with water. Many cans of concentrate have the acidity, nutrients and tannin adjusted for the production of a certain wine. Read the label to make sure you are buying the can which best suits your needs.

The possible variety of honey wines is just infinite. In addition to fruit juices you can add herbs, spices, flower petals, roots and saps. Do experiment. Every meal can become a banquet. At the same time, remember that some plants are poisonous. Make sure you are using plants recommended for kitchen use.

For the non scientific mead maker a honey or honey-fruit must probably needs additional acid if, before fermentation, it does not taste as acid as a sharp apple.

There are, of course, reliable techniques for ascertaining the acidity of wine by chemical analysis. This is fully dealt with in "Progressive Winemaking" by Peter Duncan and Bryan Acton, an Amateur Winemaking publication.

Marshalled for Melomel

TANNIN

"Skin off your nose" - Old toast

Amateur's tannin

Tannin gives to wine that necessary 'bite' or 'skin off your tongue' property one expects from a wine. It occurs in nature on the skins of fruit. Red wine is rich in tannin because this is fermented on the pulp to allow the dark skins to impart their colour to the must. Otherwise even red or black grapes produce a white juice. You have only to compare the harsh astringency of a cheap red to the soft taste of a white wine to realise the part played by tannin. Of course the great chateau wines, matured in cask and in bottle, have lost this harsh quality. Tannin is modified during the complicated alchemy of maturation.

Wines which lack tannin tend to be flat and insipid. In the old days of mead making, the rich remains of skep beekeeping went into it. Crushed larvae, pupae, pupal skins, pollen propolis and even a few crushed bees - could they have provided both yeast-nutrients and tannin? Small quantities of pollen, propolis, wax particles, honeydew and atmospheric dirt must be present, even in our present-day honey. These will make some small contribution to both flavour and fermentation. Still some additional tannin is desirable.

Stewed tea contains this substance. Old fashioned mead makers add the dregs of a good brew of tea to their must, and this is a nice, natural way of doing it. China tea drinkers should not contribute as this tea is low in tannin.

Many fruits carry tannin in stalks, pips and skins. If you are making mead when the pears are ripe, keep the skins and cores and boil them up in a little water.

You can buy tannin prepared from the skins and pips of grapes. It is a brown powder and the easiest form to use. ½-¾g (about a saltspoonful) is sufficient for 5 litres (1 gallon) of mead. One level teaspoonful weighs approximately one gram.

BASIC PRECAUTIONS AGAINST SPOILAGE

"Cleanliness is, indeed, next to godliness"- John Wesley

Keep all equipment clean. Just before use, sterilise all equipment. This may be done in one of two ways:

1) Pour boiling water over. Even glass equipment may be sterilised in boiling water, if it is first washed in hot water and detergent and hotter water used for rinsing, with a final rinse from the boiling kettle. Just as you raise the temperature slowly to avoid breakage, allow the glass to cool slowly before using it. Use lined rubber gloves to protect your hands from accidents. Pour the boiling water through the siphon tube and over your stirring spoon or sieve each time you use it.

2) Campden tablets. These are made of compressed sodium metabisulphite which, when added to a wine must, release a quantity of sulphur dioxide gas. This is the most satisfactory sterilising agent and is used by all commercial winemakers. It destroys spoilage bacteria and wild yeasts.

If the sulphite, as it is popularly called, is slightly acid, it releases its sulphur dioxide more freely. A pinch of citric acid and two tablets crushed and mixed with half a litre (a pint) of water makes a sterilising fluid. Pour it into your fermenting jar and make sure it comes into contact with the whole inner surface. Keep the solution in a screw top bottle and you can re-use it several times. Pout it through your racking tube and dip all corks before use. In Summer and Autumn, when the fruit flies are around, use it in the fermentation lock.

Campden tablets are used to keep the must healthy too, but in smaller quantity than for sterilisation of equipment. The dose is one tablet to each 5 litres (1 gallon) of wine. This should be added at the outset if unboiled fruit uices are employed. Cultured wine yeasts can do their work in the presence of a small amount of sulphite, but it is best to allow 24 hours after using the Campden tablet before the ntroduction of the yeast.

At intervals the wine must be racked. This means that the clearer wine is siphoned off the deposits which settle out. Add a tablet at each racking. This not only destroys wild yeasts and bacteria which may have entered during the time the wine was necessarily exposed to air, but will help the wine o clear. Do not drop the Campden ablet into the jar of mead. Crush it with

An infusion of over-ripe bananas gives body and helps clarity

the back of a spoon on a shallow dish. Mix with a little warm water and be sure to use every drop. Rotate the jar to ensure even distribution through the must.

PECTIN HAZE

When most fruit juice is boiled, pectin is released. This is convenient when making jam or jelly as the pectin helps to preserve the 'gel'. It is inconvenient in wine making, as it causes a stubborn haze which is difficult to clear. Juice from fruits which make good preserves

should be extracted by cold methods to avoid pectin haze. Otherwise a pectin destroying enzyme should be added. Buy this at the home wine shop and follow the instructions on the pack.

A few overripe bananas may be chopped up and boiled in about twice the volume of water required to cover them. Include one skin to every five bananas. A little of this extract added to melomels often helps them to clear and gives a little additional body.

Ready to go.

BASIC EQUIPMENT FOR MEAD MAKERS

Plastic food grade bucket with lid if possible. If no lid or if the lid does not fit tightly, cover with a clean cloth held in place by an elastic band. If the bucket becomes scratched and may harbour germs, line it with a new food-grade plastic bag.

5 litre or gallon jars for fermenting.

Half gallon Winchester bottles can often be obtained free. These make excellent storage jars.

Corks or rubber stoppers to fit the fermentation jars should have a hole bored through the centre and are fitted with a fermentation lock.

Nylon sieve.

Strainer bag. This may be bought but a remnant of terylene net curtain will suffice.

Flexible rubber or plastic siphon tube (about 1.25m 4ft.)

Long handled plastic or wooden spoon.

Bottle brush.

BASIC METHOD FOR MEAD AND MELOMEL

Day 1
Make a yeast starter as recommended on the pack.

Day 3
Dissolve honey in warm water in clean pail.
Prepare and sterilise any fruit juices or herbal infusions.
Add these to honey water.
Cover to exclude air.

Day 4
Add all other ingredients except yeast.
Stir well.
Put into warm place (20-24°C 68-75°F).
Add yeast starter.
Stir each day for 3-5 days.
Keep pail covered.

Day 7-9
Transfer contents to 5 litre (1 gallon) jar.
Insert fermentation lock.
Fill with water to within 5cm (1-2in) of top.

1 month.
Or when a firm deposit (the lees) has settled and the must has begun to clear, siphon off (rack) the must into a fresh jar. Add one crushed Campden tablet.
Top up new jar with water.
Transfer to a cooler place and ferment on.

About 3 months.
All bubbling should now have ceased and the mead should be clear. Another firm deposit should have settled. Rack again. Add Campden tablet.
Transfer to cool storage place.
The mead may need further racking as it matures.

THE PROCESS EXPLAINED

The first part of the fermentation is done in the plastic bucket so that the yeast will have access to air to assist its growth. This aerobic fermentation can be carried out in the glass jar if you wish. In this case, make the must with less water, so that the jar is only three quarters full. Plug the neck with a firm twist of cotton wool and rotate the jar each day to admit air. If you are using general purpose yeast, you add this directly to the must in the jar. During these first few days the yeast is budding and growing, using both atmospheric oxygen and that present in the must itself.

As soon as yeast growth is well established, the must (if in the bucket) is poured through a sieve standing over a funnel into the jar and a fermentation lock is fitted. The anaerobic fermentation now begins, in which the yeast works on the sugars of the must, changing it from must to wine. If the aerobic fermentation was done in the jar, fill it up now almost to the top. Do not forget to put water in your lock. You can watch the carbon dioxide gas bubbling through this water. Alcohol, equal in weight to this gas is being made inside your mead.

Do not be alarmed at the cloudy, murky appearance of the mead at this stage. As the sugars are converted, the liquid will begin to clear. In an ordinary country wine the added cane sugar (sucrose) is first converted by enzymes in the yeast into invert sugars (glucose and fructose). The bees have done that for us already. So honey musts tend to ferment more quickly than many of non honey origin.

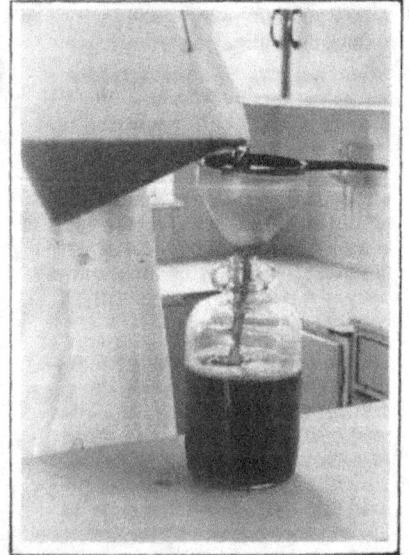
Transferring must from bucket to demi-john

Don't be alarmed it will clear

A FEW QUERIES AND PROBLEMS

RACKING

Racking is the name employed for siphoning your mead into a clean jar. It is the means of pouring off the clear must without disturbing the lees. Your wine shop will have several designs of racking tubes, some with pumps. For our purpose a length of ordinary tube will serve. Put one end into the must. Place the clean jar on a lower table or stool. Suck mead into the tube till it is almost full, nip the end between thumb and fingers and place into neck of lower jar. The mead flows gently into the new jar. Tilt the top jar to rack off the maximum amount of mead.

If you are making several gallons, pour all the lees into a bottle. This will settle out and the clear liquid can be used to fill up a jar after racking. If you are really economical some wine can be saved by filtering the lees through a filter paper. Beware of over exposure of the wine. Over exposed wine absorbs oxygen and is said to become oxidised. Sherry is a purposely oxidised wine.

ACETIC ACID

Ever present on dust particles in air are bacteria which can attack the alcohol in a wine and convert it into acetic acid. Dirty equipment, rotting fruit and especially the small fruit fly or drosophila carry the vinegar bacterium. These little black flies will appear as soon as you expose a fermenting must. So replace the lock quickly each time you rack your mead. The useful little fermentation lock allows gas to escape, but does not allow air to enter. This excludes both the bacterium and its usual carrier, the vinegar fly. Always top up the jar so as not to leave a potentially dangerous air space between the mead and the stopper.

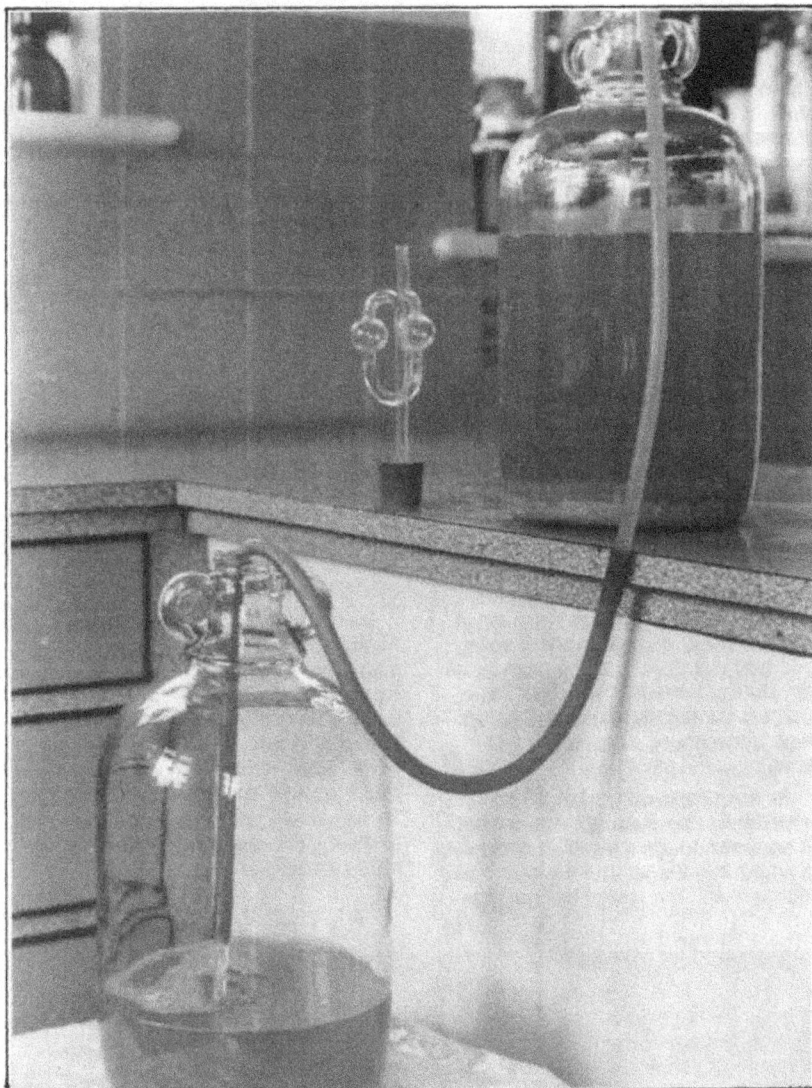

Racking

Tip the jar for last drop of clear wine

CLARITY

Occasionally you get a mead or melomel with a light flocculant deposit, which proves difficult to rack. You may also make a batch which does not clear properly.

If you have mixed your lees to preserve small quantities of precious mead, you will have noticed that this mixture settles and clears more successfully than the individual wines. You can take advantage of this characteristic by mixing wines and meads with slight problems. Not only will they usually fall clear, but you will achieve new flavours. Too dry or too acid a mead can be softened by blending with a sweet or bland one.

Blending is one of the greatest pleasures, once you have extended your mead store above a dozen gallons. It is best to store the mead in the fermenting jar till shortly before you drink it. A good idea is to take off three bottles and store the rest of the five litres or one gallon in a Winchester or half gallon jar.

Sometimes mead or honey wine stubbornly refuses to clear and you do not wish to blend it with other wines. In this case either serve it in opaque glasses or pottery goblets, or buy a packet of benonite. This is a finely divided earth which carries the lees firmly down. A very small amount is required. Follow the instruction on the pack. Proprietary clearing agents can be bought, but I have no personal experience of them.

CAPPINGS

Many beekeepers say, "I just want to wash out my cappings and make a drop of mead. Tell me simply how to do it."

This is not simple. Well drained cappings still contain a good deal of honey. I give mine back to the bees in a cappings cleaner and remove them a week later quite dry and ready for melting down for candles or recycling into foundation. I can then measure my separate honey for mead. Those who prefer to wash the honey from the cappings should place the drained cappings in a bucket and pour warm water over them. Stir very thoroughly and drain them for some time either into the settling tank or through terylene net.

With washed cappings we do not know how much honey we have in solution. The only way to find out is to use a hydrometer. This simple instrument measures the density of a liquid by the level at which it floats.
The density of our must depends mainly on the amount of sugar
in solution. It is easy to calculate the proportion of alcohol which is likely to be produced by a certain amount of sugar. This, however, is no place to go into the details of gravity and specific

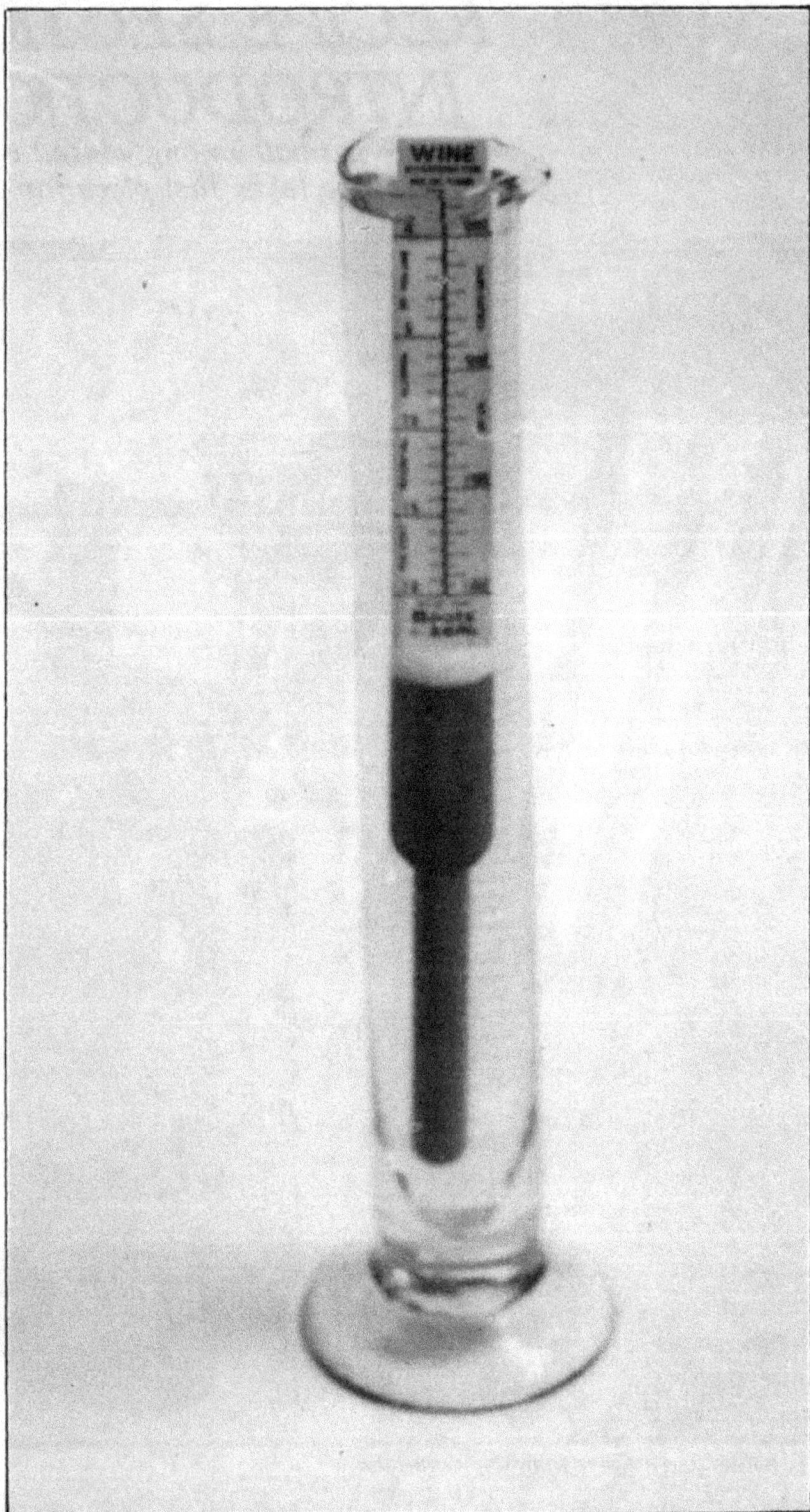

Hydrometer

gravity. Many competent authors have described the full use of a hydrometer and devotees of washing cappings are referred to them.

As a rough guide 1 kg (2¼ lbs) sugar or honey will give about 12% alcohol.

I must repeat that this booklet is for the person in his/her own kitchen making honey wines without any scientific aids. This is eminently possible. Taste the mead at each racking, learn to assess the progress of your fermentation, decide what you like and aim high. Experience teaches you to detect what is needed to produce the taste your family will enjoy. Play around with blending and train your palate on an occasional commercial tasting.

RECIPE SECTION
INTRODUCTION

"The Bee is small among winged creatures,
yet her produce takes first place for sweetness"

Six bottles and a taster from one demi-john

SMALL QUANTITIES EXPLAINED

1 teaspoonful (tspn) - as much above as below edge of spoon
½ teaspoonful - 1 level teaspoon
¼ teaspoonful - ½ level teaspoon
1 pinch - ¼ level teaspoon
1 ounce (1 oz) - 28·4 g
For gram equivalent round this number off to 30 and divide visually.

LARGE QUANTITIES EXPLAINED

Since our fermentation jars will be demijohns, recipes are intended for this quantity. They used to contain 1 gallon which is just over 4½ litres. New demijohns have a capacity of 5 litres. Recipes are given first in metric and then in imperial measures. These are not direct translations. They are rounded off to make realistic quantities in the two scales. As in all food preparation recipes must be approximations and should be varied according to individual preference.

1 can of grape concentrate refers to the size intended for making 5l (1 gal.) wine. Larger packs are available and are usually a better buy.

Since recipes are for 5l (1 gal) unless otherwise stated, add to each recipe the line: "water to 5l (1 gal)" Except where otherwise stated choose white grape concentrate.

THE RECIPES

*"One often wonders what the vintners buy
One half so precious as the goods they sell"* - Rubaiyat of Omar Khayyam
trans. Fitzgerald

MELOMELS FROM PREPARED CONCENTRATES

Many concentrates are available at home winemaking stores. They contain various combinations of fruit juices along with the necessary acids and tannin. The winemaker who wants a convenient and easy way of producing his own drinks, has a balanced must just by adding water and sugar. Substitute honey for sugar and you will add a subtle character to these wines which may now be classed as melomels.

PYMENT (I)

1 can grape concentrate
Substitute honey for sugar and follow instructions on the can. If not sweet enough for your taste, add 125g (¼lb) honey and a crushed Campden tablet and ferment on. Repeat if necessary. Rack when clear.

PYMENT (II)

½ can grape concentrate
¾kg (1¾ lb) honey
1 tspn mixed acids
pinch grape tannin
yeast and nutrient

PYMENT (III)

½ can grape concentrate
juice of one lemon
1 kg (2 lb) honey
yeast and nutrient.

SIMPLE MELOMEL

½ can grape concentrate
1 bottle apple juice (or 1 carton or fresh pressings)
juice of large lemon (or equivalent canned juice or P.L.J.)
1kg (2 lb) honey
yeast and nutrient
small pinch tannin

CYSER (I)

2 litres fresh (or two bottles or cartons) apple juice
juice of ½ lemon
¼ can grape concentrate
1 kg (2 lb) honey
pinch tannin
yeast and nutrient

CYSER (II)

3 litres fresh (or 3 bottles or cartons) apple juice
¼ tspn citric or mixed acids
1½ kg (3¼ lb) honey
pinch tannin
yeast and nutrients

Make a Melomel by adding honey.

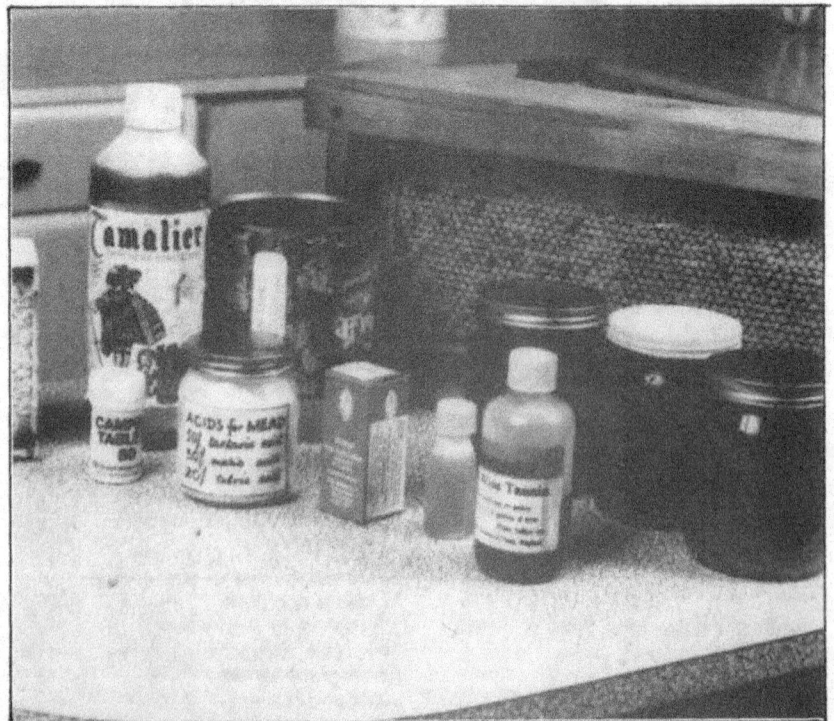
All prepared for a Pyment

CYSER (III) Economy recipe

Using above or similar recipe ferment to dryness with ½ kg (1 lb) sugar. Then add ½ kg (1 lb) honey and one vitamin B tablet and ferment on. Add more honey if still too dry for your taste.

CYSER (IV) From fresh apples

Extract the juice from fresh apples until you have about 4 litres (6 pints). Use whatever apples are to hand. A mixture of cooking and dessert apples gives suitable acidity. Fallen apples may be used.

Apple juice is most conveniently extracted with an electric juice extractor. Wooden presses and steam extractors are available.

Apples may be minced and squeezed through a strong cloth. There is no need to peel and core them. Just remove maggots. Bruises may be kept, but bad parts removed. Pour juice at once into jar and add a Campden tablet. Otherwise the juice will oxidise and go brown, as any exposed peeled apple will.

Mix 1kg (2 lb) honey with warm water to make about 1 litre (2 pints) and add to apple juice.
Leave 24 hours for sulphite to disperse and fruit particles to settle.
Pour or siphon clear juice into fermenting vessel.
Add yeast starter and ¼ tspn nutrients and proceed as directed.
This should make a medium sweet full bodied cyser. It will vary according to the sugar content of the apples. For hydrometer users an initial gravity of 100 is indicated.

For those who are fortunate enough to have fresh grape juice, and I know that many gardeners are now planting vines, make the above recipe, using half the quantity of honey for a dry wine.

APRICOT MELOMEL

1 lb dried apricots (or 1 tin)
¼ can grape con.
1 kg (2 lb) honey
1 tspn mixed acids
¼ tspn tannin
pectin destroying enzyme
yeast and nutrient

Soak apricots until soft. Chop or break up in blender.
Combine with diluted honey.
Add one crushed Campden tablet and leave for 24 hours.
Ferment on pulp for 2-3 days.
Sieve into jar and proceed as directed.

The sieved out apricot and honey solids may be boiled gently with a little sugar and served with ice cream or meringue.

PEACH MELOMEL

As for apricots, using dried or tinned or fresh fruit.

GOOSEBERRY MELOMEL

(I) Dry
2-3 lb gooseberries
¼ can grape con.
1 kg (2 lb) honey
pinch grape tannin
yeast and nutrient

(II) Sweet
3-4 lb ripe gooseberries
¼ can grape con.
1¼ kg (2½ lb) honey
pinch grape tannin
yeast and nutrient

Crush gooseberries, or break up in blender.
Mix with 1½ litres (3 pints) cold water.
Add a crushed Campden tablet and leave for 24 hours.
Mix honey with 1½ litres (3 pints) warm water.
Mix gooseberry juice and honey water. Add tannin and nutrient, then yeast. Proceed as directed.
Some winemakers would prefer to ferment on pulp for 2 or 3 days. Try both methods.

PEAR MELOMEL

If you have hard perry pears, make as for cyser.

The flesh of dessert pears does not make good wine, but the peels and cores left when preparing pears for stewing, bottling or freezing are excellent. Pour boiling water over these, macerate well and sieve into fermenting jar. Boil 3 or 4 bananas and add the infusion from these. Mix in a little grape concentrate and some sharp fruit juice such as apple if available.

Taste the must and add acid if necessary (or test for acid).

Add honey. Ferment as directed.

Taste at racking time and add more honey if you prefer a sweeter brew. This may prolong fermentation and give temporary cloudiness. Be patient.

ORANGE, GRAPEFRUIT, PINEAPPLE MELOMEL

1 large can or 2 cartons or similar quantity fresh juice
½ can grape conc.
1 kg (2 lb) honey
Infusion from one or two bananas
½ tspn acids
pinch of tannin

If using fresh juice add a Campden tablet and keep for 24 hours. Juice in cans and cartons usually contains preservative.

Mix all liquids together and proceed as usual.

RASPBERRY OR STRAWBERRY MELOMEL

2-3 lb raspberries or strawberries
⅓ can grape conc.
1 lemon or ½ tspn acids
¾ kg (1½ lb) honey
pinch grape tannin
yeast and nutrient

Macerate fruit in blender. Sieve into fermenting jar.
Add water to fruit pulp. Mix well and Sieve again. Repeat this.
Add Campden tablet and leave for 24 hours.
Add grape concentrate, acids, tannin and nutrient salts.
Adjust temperature to about 21°C (70°F)
Add yeast starter.
When vigorous ferment subsides, fill up jar and proceed as usual.

RED MEAD

Make as above with red currants.

Red currants and apples mixed make a good rosé wine.

Use a rosé grape concentrate for these two brews.

BLACK MEAD

As above, using blackcurrants and a red grape concentrate.

Add the infusion from a few bananas and some extra honey for a rich, robust dessert wine.

RHUBARB MELOMEL

Take 2-3 lb of rhubarb, chop into short pieces and place in a bowl. Sprinkle ½ kg (1 lb) sugar over. Allow to stand, well covered until sugar has leached juice from rhubarb and is all liquid. Pour liquid into fermenting jar. Cover pulp with cold water, stir well and stand aside for a time, then pour into jar. Repeat until all the sugar is dissolved and you have almost four litres (six pints) of liquid. Add ¼ can of grape concentrate and ½ kg (1 lb) honey. Ferment to dryness. Add more honey if you prefer a sweeter wine.

ROSEHIP MELOMEL

½-1 kg(1-2 lb) rosehips (small, wild rosehips are best)
¼ kg (½ lb) dried figs
¼ kg (½ lb) dried dates
Pectin destroying enzyme
Juice of large lemon and zest of skin
½ kg (1 lb) sugar
Yeast and nutrient

Pick rosehips late in Autumn, when they are ripe and red. Wash them. Crush them with a clean block of wood or place in blender with a little hot water.
Cut up figs and dates.
Pour over 2 litres (4 pints) boiling water.
Add pectin destroying enzyme, a crushed Campden tablet and the sugar.
Set aside to cool.
Add lemon, yeast and nutrient salts.
Ferment on pulp for two or three days.
Sieve out solids and ferment under lock to dryness.
Add 1 kg (2 lb) honey and ferment on.

This is a heavy dessert wine with a unique flavour and will improve with keeping.

CYSER (III) Economy recipe

Using above or similar recipe ferment to dryness with ½ kg (1 lb) sugar. Then add ½ kg (1 lb) honey and one vitamin B tablet and ferment on. Add more honey if still too dry for your taste.

CYSER (IV) From fresh apples

Extract the juice from fresh apples until you have about 4 litres (6 pints). Use whatever apples are to hand. A mixture of cooking and dessert apples gives suitable acidity. Fallen apples may be used.

Apple juice is most conveniently extracted with an electric juice extractor. Wooden presses and steam extractors are available.

Apples may be minced and squeezed through a strong cloth. There is no need to peel and core them. Just remove maggots. Bruises may be kept, but bad parts removed. Pour juice at once into jar and add a Campden tablet. Otherwise the juice will oxidise and go brown, as any exposed peeled apple will.

Mix 1kg (2 lb) honey with warm water to make about 1 litre (2 pints) and add to apple juice.

Leave 24 hours for sulphite to disperse and fruit particles to settle.

Pour or siphon clear juice into fermenting vessel.

Add yeast starter and ¼ tspn nutrients and proceed as directed.

This should make a medium sweet full bodied cyser. It will vary according to the sugar content of the apples. For hydrometer users an initial gravity of 100 is indicated.

For those who are fortunate enough to have fresh grape juice, and I know that many gardeners are now planting vines, make the above recipe, using half the quantity of honey for a dry wine.

APRICOT MELOMEL

1 lb dried apricots (or 1 tin)
¼ can grape con.
1 kg (2 lb) honey
1 tspn mixed acids
¼ tspn tannin
pectin destroying enzyme
yeast and nutrient

Soak apricots until soft. Chop or break up in blender.
Combine with diluted honey.
Add one crushed Campden tablet and leave for 24 hours.
Ferment on pulp for 2-3 days.
Sieve into jar and proceed as directed.

The sieved out apricot and honey solids may be boiled gently with a little sugar and served with ice cream or meringue.

PEACH MELOMEL

As for apricots, using dried or tinned or fresh fruit.

GOOSEBERRY MELOMEL

(I)   Dry
2-3 lb gooseberries
¼ can grape con.
1 kg (2 lb) honey
pinch grape tannin
yeast and nutrient

(II) Sweet
3-4 lb ripe gooseberries
¼ can grape con.
1¼ kg (2½ lb) honey
pinch grape tannin
yeast and nutrient

Crush gooseberries, or break up in blender.
Mix with 1½ litres (3 pints) cold water.
Add a crushed Campden tablet and leave for 24 hours.
Mix honey with 1½ litres (3 pints) warm water.
Mix gooseberry juice and honey water.
Add tannin and nutrient, then yeast.
Proceed as directed.
Some winemakers would prefer to ferment on pulp for 2 or 3 days.
Try both methods.

PEAR MELOMEL

If you have hard perry pears, make as for cyser.

The flesh of dessert pears does not make good wine, but the peels and cores left when preparing pears for stewing, bottling or freezing are excellent. Pour boiling water over these, macerate well and sieve into fermenting jar. Boil 3 or 4 bananas and add the infusion from these. Mix in a little grape concentrate and some sharp fruit juice such as apple if available.

Taste the must and add acid if necessary (or test for acid).

Add honey. Ferment as directed.

Taste at racking time and add more honey if you prefer a sweeter brew. This may prolong fermentation and give temporary cloudiness. Be patient.

ORANGE, GRAPEFRUIT, PINEAPPLE MELOMEL

1 large can or 2 cartons or similar quantity fresh juice
½ can grape conc.
1 kg (2 lb) honey
Infusion from one or two bananas
½ tspn acids
pinch of tannin

If using fresh juice add a Campden tablet and keep for 24 hours. Juice in cans and cartons usually contains preservative.

Mix all liquids together and proceed as usual.

RASPBERRY OR STRAWBERRY MELOMEL

2-3 lb raspberries or strawberries
⅓ can grape conc.
1 lemon or ½ tspn acids
¾ kg (1½ lb) honey
pinch grape tannin
yeast and nutrient

Macerate fruit in blender. Sieve into fermenting jar.
Add water to fruit pulp. Mix well and Sieve again. Repeat this.
Add Campden tablet and leave for 24 hours.
Add grape concentrate, acids, tannin and nutrient salts.
Adjust temperature to about 21°C (70°F)
Add yeast starter.
When vigorous ferment subsides, fill up jar and proceed as usual.

RED MEAD

Make as above with red currants.

Red currants and apples mixed make a good rosé wine.

Use a rosé grape concentrate for these two brews.

BLACK MEAD

As above, using blackcurrants and a red grape concentrate.

Add the infusion from a few bananas and some extra honey for a rich, robust dessert wine.

RHUBARB MELOMEL

Take 2-3 lb of rhubarb, chop into short pieces and place in a bowl. Sprinkle ½ kg (1 lb) sugar over. Allow to stand, well covered until sugar has leached juice from rhubarb and is all liquid. Pour liquid into fermenting jar. Cover pulp with cold water, stir well and stand aside for a time, then pour into jar. Repeat until all the sugar is dissolved and you have almost four litres (six pints) of liquid. Add ¼ can of grape concentrate and ½ kg (1 lb) honey. Ferment to dryness. Add more honey if you prefer a sweeter wine.

ROSEHIP MELOMEL

½-1 kg(1-2 lb) rosehips (small, wild rosehips are best)
¼ kg (½ lb) dried figs
¼ kg (½ lb) dried dates
Pectin destroying enzyme
Juice of large lemon and zest of skin
½ kg (1 lb) sugar
Yeast and nutrient

Pick rosehips late in Autumn, when they are ripe and red. Wash them. Crush them with a clean block of wood or place in blender with a little hot water.
Cut up figs and dates.
Pour over 2 litres (4 pints) boiling water.
Add pectin destroying enzyme, a crushed Campden tablet and the sugar.
Set aside to cool.
Add lemon, yeast and nutrient salts.
Ferment on pulp for two or three days.
Sieve out solids and ferment under lock to dryness.
Add 1 kg (2 lb) honey and ferment on.

This is a heavy dessert wine with a unique flavour and will improve with keeping.

SPICED MEAD OR HIPPOCRAS

*"The righteous shall be given to drink
pure mead sealed with musk" - The Koran*

Some meads may not be so palatable as others. We can't realise our full hopes all the time.

From time immemorial spices and condiments were used to mask unpalatable tastes, and - let's face it - to make marginally acceptable those foods and drinks which had succumbed to their natural enemies of deterioration and corruption. We are so well aware of the pitfalls, that, if we observe proper hygienic rules, we shall not meet any of the wine diseases in our meads. Still there may be better batches than others. It may be that some are just too dry for our palate. In that case try adding a little well flavoured honey. If the sweetness is acceptable but the flavour doesn't appeal, choose those meads and melomels as a base for hippocras.

Take whole spices from the store cupboard -the sort which are ground up for mixed spices. These include cloves, nutmeg, mace, cinnamon stick, peppercorns, coriander, ginger. Orange, grapefruit and lemon peel can be sliced very thinly (wash the fruit first) and dried in a cooling oven after baking. This makes a good, aromatic flavouring for mead. Drop small amounts of these ingredients into a jar of mead. If you have only the powdered form, make a small bag of muslin or cambric or curtain net and suspend it on a piece of cotton in the mead. Use a Winchester or a large wine bottle for various experiments.

Taste the mead every week and when sufficient flavour has been leeched from the spices, rack the clear hippocras into bottles. Add an extra pound of honey and serve hippocras piping hot at a Winter party. Wine spices may be bought in individual portions in sachets similar to tea bags. Just suspend the bag in your bottle - or in a glass of hot, sweet mead - for a splendid "gluh-wein".

HERBAL MEAD OR METHEGLIN

*"**Biron** White-handed mistress, one sweet word with thee.
Princess Honey and milk and sugar, there is three.
Biron Nay, then two treys, - and if you grow so wise,
Metheglin, wort and malmsey, well run dice.
There's half a dozen sweets" - Shakespeare
Love's Labour's Lost
Act 5 Sc 2*

All the popular culinary herbs have a place in the making of metheglins, which are useful to serve as aperitifs.

They may be gathered fresh from your own garden or bought dried at the herbalists. Selections of aromatic herbs were used in ancient days by monastic wine makers. They were collectively called gruit, and the brothers tilled their gruit gardens for their production. Famous liqueurs, whose flavour depends on the subtle qualities of the gruit employed, have survived to the present day. Bitter herbs determine the taste of many famous aperitifs. Fennel helps to flavour the ouzo of Greece.

Wormwood is the main ingredient of the wide mixture used in vermouth. Packets of these gruit mixtures may be bought at wine stores or herbalists. They are suspended in the mead you wish to flavour.

It is fun to experiment with your own fresh herbs. Sprigs of thyme, rosemary or fennel standing in a bottle of crystal clear mead or melomel makes a lovely gift. The flavour gradually enters the liquid, so you should taste it to make sure you have not overdone the herbal quality. You can with advantage fortify these offerings with a little brandy, vodka or other spirit. This will inhibit any tendency to disease in the mead from introduction of the unsterile plant material.

Try mixing herbs and spices. Make a white; a rose and a red; a dry, a medium and a sweet. The variety has no end.

Many country folk have faith in certain herbal remedies. These tissanes made into honey wines or mixed with mead are an excellent way of taking medicine. The bitterness of the herb and the sweetness of the honey make a dual taste of charm. They may be mixed with soda water, tonic or bitter lemon to make a long drink.

HEATHER MEAD

Final racking

The strong aromatic flavour of heather honey is preserved in the drinks made from it. I have little experience of making heather meads and honey wines becuase I do not live in a heather district. I have used a proportion of heather in a few meads, along with mild flavoured honey. These drinks had a distinctive savour. Heather gives a fine aroma to mead, so a small proportion in a batch intended for showing would provide extra bouquet.

Heather honey is reputed to be used in Bonnie Prince Charlie's liqueur, Drambuie. Heather mead certainly makes a rich hippocras. Perhaps our Yorkshire and Scottish beekeepers have more experience of this tangy brew.

DANDELION METHEGLIN

Dandelion flowers
1 kg (2 lb) honey
⅓ can grape concentrate
a little apple juice if available
juice of 2 lemons (or 1½-2 teaspoons of acid)
¼ teaspoon tannin
yeast and nutrients

Gather dandelions on a warm, sunny day. Discard the green calyx and stem, retaining only the yellow petals. The quantity is not vital, but you can go up to two litres of petals (not pressed down). Pour boiling water over petals and stir. When cool sieve out petals, add other ingredients and proceed as directed.

COLTSFOOT METHEGLIN
GORSE METHEGLIN
MARIGOLD METHEGLIN
WALLFLOWER METHEGLIN

As above. These flowers are not as prolific as dandelions, so quantity may be reduced.

I do not include cowslips in my recipes, though this was a most popular wine in days gone by. Cowslips are now a protected species. You could, of course, cultivate your own.

ELDERFLOWER METHEGLIN

As above, but use only 3-5 heads of flowers. This has a very pungent taste, much overused by home winemakers. It is good when delicate. I am often given a bottle of elderflower wine and I blend it with a gallon of mead. I find this is sufficient to flavour six bottles of metheglin.

Any tissance or tea made from a culinary herb makes metheglin. I specially recommend parsley. Use only the green leaves, not the stems which produce a slimy mucilage.

HONEY FOLLY

Collect vine prunings only from vines which have not been sprayed. Use only soft green parts. Proceed as for flower metheglins.

ROSEPETAL METHEGLIN OR RHODOMEL

This is a particular favourite of mine. It may be made as a white, rose or red wine. Pick off the petals just before they fall on a warm, sunny day. Use white and yellow roses with a white grape concentrate; choose dark red, strongly scented petals and a red concentrate together. A mixture can give a pink or rose wine. Slip in any oddments of fruit which happen to be ready at the time - white or red currants, raspberries and especially the juice from the early fallings from the orchard. The white can be a light table wine, medium sweet and delicate. The rose should have a similar sweetness and a light and fragile pink colour. (This is where a handful of rapberries are invaluable). The red should be a heavy, rich, sweet dessert wine with a strong aromatic bouquet. Add the infusion from a few bananas and use a port yeast.

A handful of rosepetals, marigolds, elderflowers added to any mead or melomel in season will enhance the bouquet.

The water used for blanching beans for freezing may be used when making honey wines. I find that wines made with this water clear very readily.

HONEY BEER

"All hail, thou big and foaming bowl,
Hail constant idol of my soul;
How laughingly the bubbles rise
Upon thy rich and sparkling tide" Brasenose College Verse

n his diverting "Wassail in Mazers of Mead" (1948) G. R. Gayre says, "The knowledge of malting spread from the ancient, warm, temperate world to the North, and the use of malted barley (as a cheap substitute) took the place of honey". He is arguing that all early beers and ales were a fermentation of honey with herbs for flavouring. This is substantiated by a much quoted reference to an adventurous Greek explorer of the fourth century B.C. named Pytheas. He kept a diary of his journeys, in which he stated, "Wherever grain and honey flourish men use it for making drink".

Primitive peoples of today often show what must have been the behaviour of our own ancestors. We can probably draw some conclusions about the early production of honey ale in our own islands from reading accounts of that made by the Mbeere of Kenya (Bee World Vol 53 No 3 1972 p 121).

The Anglo-Saxon word for a feast or banquet was a 'beer-drinking' (ge beor scype). The early editions of the Bible translate sicera as beor. Surely this word comes from beo = a bee, and indicates that the venerable Bede was copying a Bible translated by a people whose 'strong drink' was honey ale.

The mead of Valhalla was undoubtedly of beer strength. This is obvious from the yeasts available to Scandinavian heroes of old. You can tell it was a long drink from the size of the horns which were their beer mugs. This honeyed ale would come in as many varieties as does our present day beer. Flavourings and additives to produce body or assist fermentation would depend on the season and the vegetation of the region. Fruit juices, maltings and grain, flowers and saps, roots, herbs and spices would all be tried. The Picts and Scots are said to have added heather blooms to their honey ale.

The yeast, called barm or godesgood, since its function seemed so mysterious, was saved from each good fermentation to start another. An old law enjoined each brewery to conserve some of this bounty to give to honest citizens requiring to ferment their home brew. This has never been repealed and you can still take your (sterilised) screw top jar to any brewery and beg a small amount of 'godesgood'. Most of them keep a charity box so that you can pay for this bounty.

Up to the Great War home brew was drunk by the children and servants and other less privileged members of rural households as soon as it began to ferment. The old country name 'barley broth' supports this.

Good old English ale of Mediaeval times had lost its connection with honey and was made of malt, water and yeast. It is certain that many bitter herbs were used in all home brews, both for flavour and as a preservative. Yet, when hops were used by immigrants from Europe, in the fifteenth century, they were viewed with much suspicion by brewers and drinkers alike. Petitions were made to Parliament to limit their use, and much rivalry ensued. Little by little, however, the keeping qualities of the hopped ale and its refreshing, bitter taste made it the most popular drink. Before the days of tea and coffee ale was drunk by all Englishmen. The nourishing qualities of this healthy beverage was recognised by continued legislation - the assize of bread and ale, which kept down the price of these two staples.

The bastions of brewing were the monasteries. If one were to believe the accounts preserved in some of them, monks and nuns must seldom have been sober. Since many religious houses also kept bees, it is fairly sure that malt and honey would often be mixed to produce honey ale. Many modern recipes for mead, especially

A mead mazer

those printed in American beekeeping journals, include both hops and grain. They are really strong beers rather than wines and are comparable with barley wine.

Honey ale has an advantage over mead in that the thirsty beekeeper can be broaching it a month after making it. It is also less complex and we can rely on good results every time, provided the usual rules for cleanliness are observed.

APIARY ALE for beginners in home brewing.

Buy a beer kit. Make it up according to the instructions on the container, but substitute honey for the sugar in the recipe.

HONEY ALE A reliable recipe for 10 litres (2 gallons)

Put into a scrupulously clean bucket of at least two and a half gallons capacity:

500g (1 lb) malt extract
1 kg (2 lb) honey
½ tspn salt
⅛ tspn yeast nutrient
Juice of half a lemon

Put into a bag made from muslin or terylene net:

250g (½ lb) crystal malt (broken with a hammer or run for a few seconds in an electric blender)
30g (1 oz) hops
a handful of barley or rice flakes

Boil bag and contents in largest available pan for ¾ hour.
Pour the infusion over the malt, honey etc.
Fill up the pan with hot water and do a second short boil to get maximum

extraction of the grains.
Make up to 10 litres (2 gallons) with tap water.
Wait until it has cooled to 24°C (75°F). Add beer yeast.
Stir well. Keep at room temperature.
If a thick scum forms on the surface, remove with a fish slice.
When ale begins to clear, (5-7 days) rack into clean pail.
Discard yeast deposit. Wash the bucket thoroughly.
After 2-3 days it will look almost clear. Rack into original bucket.

Add 40g (1½ oz) sugar dissolved in a little hot water. Stir.
Fill immediately into clean beer bottles, leaving about 25mm (1 inch) gap.
Stopper firmly or add crown corks.
Keep for three weeks. It will improve if kept two months.

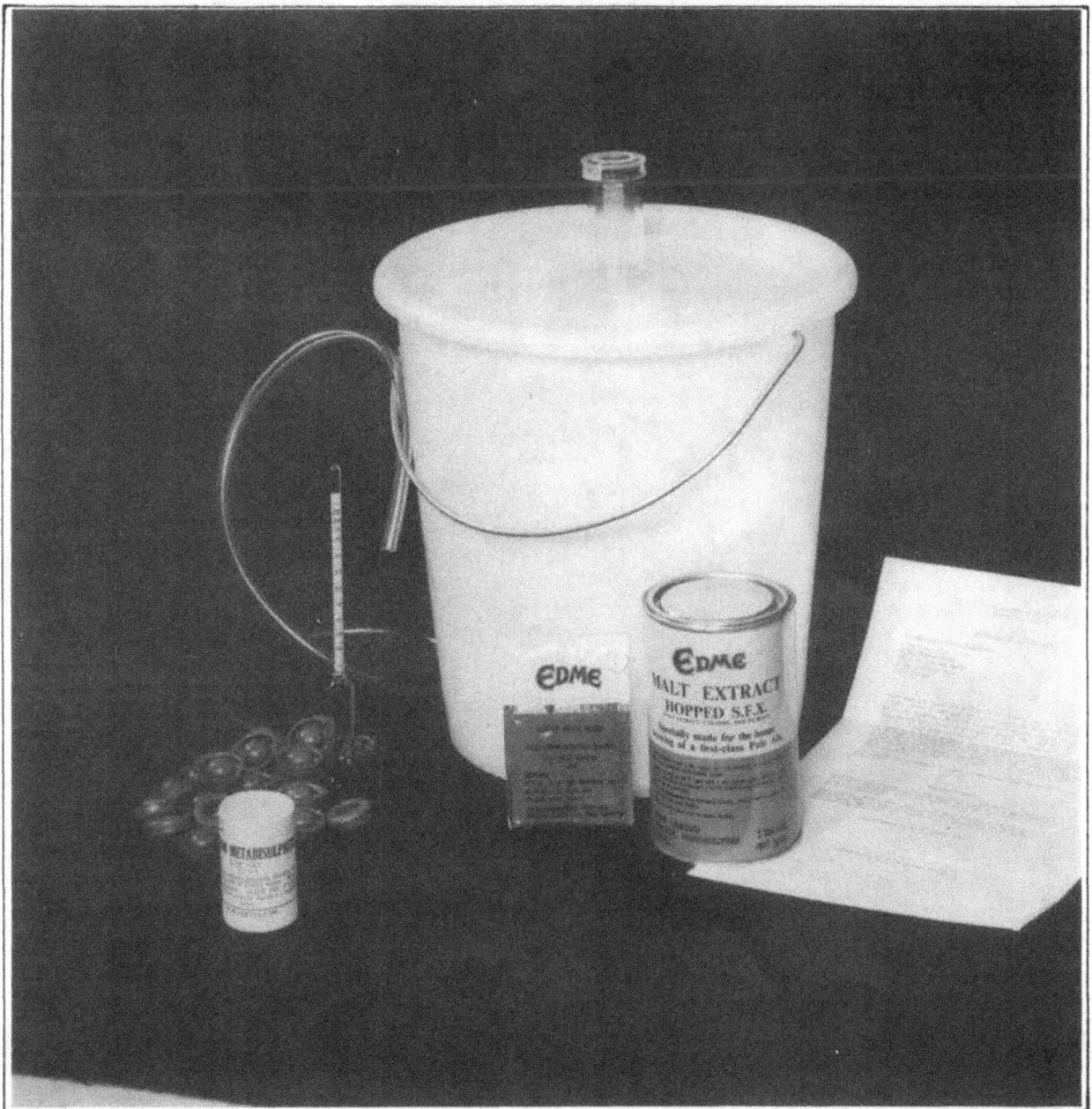

Ready for a batch of Honey Ale

BEEKEEPER'S BARLEY MEAD Recipe for 10 litres (2 gallons)

1¾ (3½ lb) pale malt, crushed
125g (4 oz) flaked barley or rice or oats or maize
65g (2 oz) hops
1 kg (2 lb) honey
2 level teaspoons gypsum
2 level teasponss citric acid
1 teaspoon salt
2 Campden tablets
Champagne yeast and nutrients

Put pale malt and flakes through the mincer or crush with a rolling pin or place in blender for just a few second. Whiever method you choose, break it up, but not too finely. This is called the grist.
Place this in your largest pan.
Heat some water to 68°C (152°F)
Pour over grains till pan is almost full.
Keep at this temperature for two hours.
Strain solids out. This liquor is called the wort.
Return wort to pan, add gypsum and hops, boil for 40 to 45 minutes.
Place all other ingredients except yeast starter into clean fermenting bucket.
Strain contents of pan into bucket and stir well.
When wort has cooled to 24°C (75°F) add yeast starter. Stir well.
Cover with lid or cloth. Skim off any thick scum which forms.
After 3 or 4 days transfer to two fermenting jars and fit fermentation locks.
Rack after about a month to six weeks when deposit is firm and mead begins to clear.
Ferment on until quite clear. Rack into clean jar.
Add 20g (¾ oz) sugar to each jar and bottle at once.
Small non returnable screw top bottles are good for barley mead.
Barley mead is served in wine glasses and not in pints.
It may be drunk a month or six weeks after bottling but improves if kept longer.

BEEKEEPER'S STOUT (5 litres - 1 gallon)
125g (4 oz) whole black malt
60g (2 oz) flaked barley
125g (4 oz) crushed crystal malt
250g (8 oz) malt extract

Place these ingredients in a pan with 2 litres (4 pints) of water.
Heat to 65°C (150°F) and hold at this temperature for ¾ hour.
Tie 20g (¾ oz) fuggles hops in a muslin or terylene bag.
Strain wort over hops and simmer for 1½ hours.

Put into clean fermenting bucket:
½ teaspoon citric acid
½ teaspoon salt
½ teaspoon yeast nutrient
500g (1 lb) honey

Remove bag of hops and pour hot wort over these ingredients
Stir to dissolve. Make up to 5 litres (1 gallon) with tap water.
Cool to 24°C (75°F). Add beer yeast.
Proceed as for honey ale.

HONEY NETTLE BEER (5 litres - 1 gallon)

Choose young green nettle shoots. Use rubber gloves to pick them. The quantity is not vital but the shoots, not pressed down, could fill your 5 litre (1 gallon) bucket.

125g (4oz) crystal malt (broken)
500g (1 lb) malt extract
250g (½ lb) honey
Piece (about 15g or ½ oz) bruised ginger
Juice of lemon or 1 tspn citric acid
¼ tspn salt
Small handful dried hops
Ale yeast

Simmer nettles and crystal malt in large pan for about 40 minutes.
Five minutes before end of boiling toss in small handful of hops.
Put all other ingredients into clean fermenting bucket.
Sieve contents of pan on to these ingredients and stir.
Squeeze nettles to extract all water.
Make quantity up to 5 litres (1 gallon) with tap water.
Cool to 21°C (70°F). Add yeast.
Proceed as for honey ale.

DANDELION HONEY BEER

Dig up two dandelion plants - roots, leaves, flowers.
Wash and remove any unwholesome parts.
Use these in place of nettles in above recipe.

FOR THE CHILDREN

Many good unfermented drinks can be made with honey. Honey lemonade is refreshing on a hot day. Honey, glycerine and lemon is a famous cure for colds. An infusion of dried elderflowers and peppermint, sweetened with honey and taken hot at night is a good treatment for a head cold. Any herbal remedy can be made palatable in this way. Honey in hot milk is a universal nightcap. Here are a few other suggestions:

HYDROMEL

Combine honey, water and any fresh or tinned fruit juice for a refreshing and invigorating drink.

HONEY LEMON CORDIAL

1 large lemon
30g (1 oz) tartaric acid
500g (1 lb) honey
1 litre water

Combine all ingredients, having water at no more than 70°C (150°F) so as not to spoil honey. Keep in refrigerator. Dilute with water or soda water.

HONEY GINGER BEER

Keep a dessertspoonful of yeast from the beer making OR beg it at the local brewery OR work 15g (½ oz) baker's yeast with a teaspoonful of sugar.

Add:
2 cups water
1 teaspoon sugar
1 teaspoon ground ginger

Plug neck of bottle with cotton wool or keep in screw top jar with lid not quite screwed down. This is your ginger beer plant. Keep in for a week, feeding it every day with a teaspoon of sugar and ¼ teaspoon ginger.

Strain through muslin into large basin or bucket.

Add:
Juice of three lemons
2 cups sugar
1 cup honey
4 cups hot water

Stir until dissolved. Add 16 cups cold water. Bottle the mixture, leaving about 50 mm (2 in) space at top of liquid. It is ready to drink in a week. The 'beer' is drunk whilst still fermenting. Keep in a very cold place or drink up your ginger beer. It will blow the corks or break the bottles if kept too long in warm conditions. Never use a screw stopper. It is not funny to have a bottle blow up when you are opening it.

Divide what is in the cloth into two jars. Add two cups of water to each jar. Give one away and begin to feed your own plant for another week, when you make another batch. This is a good hot weather game and soon your whole neighbourhood is supplied with honey ginger beer plants.

This may be made with sugar instead of honey by your non beekeeping friends of course.

ELDERFLOWER HONEY FIZZ

Place in a clean bucket:
5 litres (9 pints) cold water
rind and juice (no pith) of two lemons
1 kg (2 lb) sugar
9 heads of elder flower
2 tablespoons honey
1 cup mead or 2 tablespoons white wine vinegar

Stir contents daily until you see bubbles rising.

Bottle liquid through a sieve, leaving about 30 mm (1 in) space. Drink after three days. Keep any fizz you do not drink in a very cold place, or it will become too effervescent and may blow the corks. Do not keep in screw top bottles or there is a danger of burst bottles.

HONEYED MILK SHAKES

Place in a blender 1 banana, 1 tablespoon honey and a glass of milk. Blend for a few seconds. Use any other fruit for variety.

MULLS, PUNCHES AND OTHER MIXES

Hot Summer days call for garden meetings where beekeepers ostensibly learn about each other's apiaries, though most of them are there for the refreshments. Soon there's hallowe'en. Cold Winter brings the big festivals of Christmas and New Year, not to mention the association party. Warming cups cheer the dark season and thaw frosted fingers and toes. These are the times when hot metheglin or hippocras can become the life and soul of the party. Our not-so-good meads and metheglins can be used to mix punches and fruit cups. The quantities given are approximate. They are intended merely as a guide line: be bold and enjoy yourselves.

Punch, they say, comes from a word in one of the Indian languages meaning five. These are sweet and sour, strong and weak + your basic wine. If it is mulled it is sweet and hot. With ice, punch can be a long, cold refreshing experience.

APIARY CUP

Slice half a cucumber very thinly. Place in a punch bowl and squeeze a lemon over it. Pour 100g (2-3 oz) honey over this and a bottle of sweet mead or melomel. When the honey is incorporated, add a bottle of soda water. Cool in an ice bucket and add ice to each glass.

SWARM CONTROL SLING

1 bottle very sweet mead
1 bottle red melomel
1 glass of a liqueur which will match the melomel
1 glass brandy

Serve with crushed ice and a slice of lemon on each glass.

DRONE STINGO

1 bottle honey ale
1 wineglass strong sweet mead
Peel and juice of half a lemon
small pinch nutmeg

Serve on ice.

CYSER CUP

1 bottle cyser
1 bottle honey ginger ale
small glass vodka

GARDEN MEETING CUP

1 bottle of any melomel
2 bottles elderflower fizz
juice and rind of 1 orange and 1 lemon
Slices of orange, lemon and an unpeeled apple
Thin slices of cucumber
2-3 small sprigs of mint and/or borage

Chill melomel and fizz in refrigerator and mix with other juices.
Float slices of fruit, cucumber and herbs on surface.
A little gin or vodka will give it more kick.
Fizzy orange or lemonade with make this into a long, garden party drink.

HONEY FRUIT PUNCH

1 tin fruit salad or any fruit in season
1 bottle honey lemonade (or diluted honey/lemon cordial)
1 bottle honey gingerbeer
1 bottle honey elderflower fizz
cracked ice

Place chosen fruit in punch bowl. Add cracked ice.
Add fizzy drinks just before serving.
Serve with spoon and straw to each tall glass.
For a children's party a scoop of ice cream may be added to each glass.

SWEET MEAD AND LEMON MULL

Stick 8-10 cloves into a lemon
Place in a fairly hot oven for 15 minutes
Warm about 100g (3 oz) honey in a punch bowl
Place hot lemon on honey
Heat mead in a stainless steel pan and pour over honey and lemon.
Serve really hot.

MELLIFERA MULL

1 bottle metheglin or hippocras
a little lemon juice
1 tot whisky
4 cloves
¼ tspn mixed ground spices

Heat in a stainless steel saucepan. Pour into warmed glasses through a fine sieve.

LAMBS WOOL

Core several small apples. Bake them in the oven until the flesh begins to ooze through the slits in the skin.

Place in a punch bowl 3 or 4 pieces of crystallised ginger cut very small. Grate nutmeg coarsely over this and pour a little warm honey over. Heat two pints of honey ale until it is a good temperature for drinking. Pour into bowl. Stir well, then place the hot apples on top. This is a good drink before going for a Winter walk or to midnight mass.

BEES' WASSAIL

2 bottles honey ale
1 bottle sweet honey wine
250g (½ lb) honey
Apples
Nutmeg, cinnamon, cloves, ginger
Lemon and/or orange

Cut apples into slices and poach them gently.
Grate rind of well washed lemon and/or orange into punch bowl.
Add spices and honey.
Heat honey ale and wine. Pour into bowl. Float poached apples on top.
Remove pith from lemon and/or orange and arrange slices round bowl.
Ornament bowl with black paper cats and witches for Hallowe'en or with holly and mistletoe for Christmas.

HONEYFLOW GLUHWEIN

Buy packets of gluhwein spices from the home wine store.
Heat a less good mead or melomel.
Serve in narrow glasses with a sachet of spices and a slice of lemon to each glass.

HARVEY'S SPECIFICK

(He discovered the circulation of the blood. This helps it!)

1 glass brandy
Juice of one lemon
3 tablespoons honey
1 bottle mead or melomel

Drink hot or cold.

HONEY LIQUEUR

Put equal quantities of brandy or vodka and mead into a bottle.

Add cloves, nutmeg, mace, chopped preserved ginger and a stick of cinnamon and set aside. Taste occasionally and when sufficient flavour has been extracted, sweeten to taste with heather honey. Set aside to clear.

COFFEE*ORANGE*HONEY*LIQUEUR

Push 16 coffee beans into the skin of a well washed orange.
Place this in a tall jar (Kilner or coffee jar).
Cover orange with spirit (vodka, brandy or Polish spirit)
Pour over 250g (½ lb) dark honey
Top with mead, using similar quantity to spirit. Screw down lid.
Turn jar to assist mixing every now and then. Leave for six months.
The orange will shrink and dry. The resultant liqueur is out of this world.

Several honey liqueurs may be made as above, using figs, cherries, sloes, blackberries, damsons - almost any fruit.

Prepared liquid essences may be bought at home wine shops. Prepare as directed, substituting honey for sugar in recipe.

JUDGING AND TASTING MEAD

"For now I'm a judge, And a good judge, too"- Gilbert

It is disconcerting when you have made a real "vintage" mead to have your guests toss it off as though it were a cola. Learn to treat your honeyed wines with the respect due to great achievements. Serve them with distinction in the proper glasses and match them with care to the food you choose to accompany them. The proper glass for any wine is simple, tulip-shaped and stemmed. If the mead is light, white and chilled, you can hold the glass by its foot or stem to avoid warming the wine with the hand. A heavy red melomel can be cupped in the hands to warm the wine and release more bouquet. You may even warm chilled fingers on the hot punch served to begin a Winter party.

THE APPEARANCE

The first thing to appreciate in your mead is its appearance. Hold your glass up to the light and see its rays passing through the crystal clear liquid and dancing on its surface. If the mead is slightly effervescent, watch the bead rising from the glass on to the surface and enjoy its constant re-creation.

Swirl the mead in the glass and watch how it clings. If it is rich in alcohol, it will leave waves clinging to the glass, which descend in gentle falls back into the mead. It will also release what we go on to enjoy next.

THE BOUQUET

Now put your nose right into the top of the glass and breathe in gently and deeply. You should be regaled by a vision of all the essences of the blooms visited by the bees collecting the nectar which went to make the honey for your mead. Shut your eyes and conjure up those honey-flow days. Your melomels will have in addition the aroma of whatever fruit or herb went to its making. Where you have a handful of rose petals or a few elderflowers, their fragrance will filter through.

Dry or medium mead should be a light wine and the nose should not detect the smell of alcohol. In a very young mead or melomel, the bouquet will be too raw and too redolent of the separate ingredients. Let your meads mature to enjoy the full blending of the bouquet. Repeat the breathing and, as these mixed perfumes are borne into your nose on the volatile exudances of your mead, prepare for the third pleasure, that of taste.

THE TASTE

There are four basic taste areas. At the tip the tongue seeks sweetness. As the drink passes over the surface, the tongue detects saltiness and a large range of pleasures included in this field. Rolling over the edges of the tongue the acidity of the mead is felt. Finally, at the base of the tongue, just as you are swallowing the drink, its bitterness becomes evident.

All these tastes must be given time if we are really to savour mead, and especially if we seek to judge it in comparison with others. Try the process in private at first. It isn't easy. Take a mouthful of mead; feel it on your lips and the tip of the tongue as it enters your mouth. This first attack of the taste lasts so short a time - only a few seconds. Your mind must be prepared to receive it.

Now allow the liquid to lie in your mouth and take a breath through your nose. Roll the mead over the surface of your tongue, between the teeth and along the palate. Draw air into your mouth through the mead and continue to roll it over every surface with a chewing action. It may sound a little like gargling, but this is quite acceptable. Exhale the breath through your nose.

At this stage judges spit out the bulk and swallow only the tiny amount left in the saliva. Those just enjoying it swallow the mead and savour the after taste.

The total of these experiences of eyes, nose and mouth should be a pleasing balance of sensation. That is why we took care with the honey, the acid, the tannin and the hygiene.

If judging wines you will need to refresh your palate. Cheese, small plain or savoury biscuits, small chunks of bread or glasses of water are popular. If enjoying mead with a meal or as an accompaniment to an evening's chat, by all means make a thorough tasting at one point, but do not press the exercise upon guests. Nevertheless it is good to hold the mead in the mouth long enough to allow the whole inner surface to savour it.

The Mature Mead Maker

"I am as sober as a judge"
Henry Fielding - Don Quixote in England III(4)

In time your palate will become more sensitive. You will differentiate between dryness and sweetness and the full range between. Your preference will gradually move towards the dryer meads and you will enjoy the tiny residual flavours of the honey you used. At the same time you will learn to assess the proportion of sweetness needed in a heavier dessert or social mead. You will relate this to the higher proportion of alcohol which may be permitted and the robust body of this type of wine. As awareness of the lack of excess of alcohol, acidity, astringency, sweetness, bitterness, body and texture matures, you will modify the recipes to suit your palate.

As you pay attention to what before may have seemed lesser aspects of mead: the "robe" or appearance, the "nose" or bouquet, blending becomes important. Make sufficient batches to be able to experiment with blending. You will be astonished at your achievements. Meads, metheglins and melomels with disappointing imbalance can so complement one another, that two or three of these can be combined into a stable, clear and satisfying whole.

MEAD TASTING

*"When the pub is sighted
In the market square
Every face is lighted
With its rosy flare.
Then says every cheery soul,
Could you find a better hole?"*

*Carmina Burana (12th Century)
trans: Helen Waddell.*

May I recommend a plan for developing mead-expertise? During the long off-season, when the bees are soberly clustered, invite a group of mead-loving friends for the evening. Provide palate cleansers by way of small hunks of cheese, tiny plain biscuits, bland nuts. Wrap each bottle in a napkin to disguise its origin, (but keep a record). Give the guests a sheet of paper with headings or columns labelled: appearance, bouquet, sweetness, acidity, astringency, flavour, balance, type of honey used, fruit, flowers, herbs etc. (or a selection of these).

Take your time. This can't be rushed. Be honest. Meadmakers want to learn and correct any errors. Good beekeepers won't be put off by a measure of criticism. At the same time try to make positive remarks. Suggest improvements rather than condemn faults.

As the group develops, have evenings devoted to commercial wines. A claret, a Loire, a Rhine, a Mosel, a Spanish or Italian evening. If each pair of guests brings a bottle, the prices car be pooled. Experience of true wine serves to improve appraisal of our own mead.

Presentation of Mead for shows

"Hire mouth was sweet as bracket or the meth"
Chaucer Miller's Tale

To succeed on the show bench, mead must be quite clear, free from 'floaters' and presented in the correct bottle. Read the schedule carefully and comply with all the requirements for bottle shape, type of cork and positioning of label. Twist the bottle rapidly and, if any small specks are seen to rise from the base and float about in the mead, discard the bottle or rack it off. Make sure there is the correct space between the surface of the mead and the bottom of the cork. If the cork tends to rise, insert a short piece of thin string along with the cork, then slowly withdraw it. Any air compressed by the cork will be released. Be sure you have entered the right class for sweetness and content.

Wipe the bottle with a clean, lint-free duster sprinkled with methylated spirit and polish its surface. Do not get any on the cork and do this early enough for all smell of spirit to be well away before the judge gets at it.

At all judging or tasting avoid any extraneous scent or taste which may have an adverse effect on judgement. Ladies should not wear perfume or perfumed makeup. Gentlemen should avoid scented after-shave. The washrooms should contain unscented soap. And, of course, smoking is unthinkable. If, in a small show, the same judges for honey and mead must be employed, the mead judging should precede the honey. Dry mead classes must be judged before those for sweet mead. A dry mead can not be appraised by anyone who has been tasting honey all morning.